# DARWINISM

TWAYNE'S STUDIES IN
INTELLECTUAL AND CULTURAL HISTORY

**Michael Roth, General Editor**
*Scripps College and the Claremont Graduate School*

Previously published

Anarchism
*Richard D. Sonn*

Dissent and Order in the Middle Ages: The Search
for Legitimate Authority
*Jeffrey Burton Russell*

The Emergence of the Social Sciences, 1642–1792
*Richard Olson*

Liberalism Old and New
*J. G. Merquior*

Renaissance Humanism
*Donald Kelly*

# DARWINISM

Peter J. Bowler

Twayne Publishers   •   New York
Maxwell Macmillan Canada  •  Toronto
Maxwell Macmillan International • New York  Oxford  Singapore  Sydney

*cop1*

*Darwinism*

Twayne's Studies in Intellectual and Cultural History, No. 6

Twayne Publishers
Macmillan Publishers Company
866 Third Avenue
New York, NY 10022

Maxwell Macmillan Canada, Inc.
1200 Eglinton Avenue East
Suite 200
Don Mills, Ontario M3C 3N1

Macmillan Publishing Company is part of the Maxwell Communication Group of Companies.

**Library of Congress Cataloging-in-Publication Data**

Bowler, Peter J.
    Darwinism / Peter J. Bowler.
        p.      cm. — (Twayne's studies in intellectual and cultural history)
    Includes bibliographical references (p.   ) and index.
    ISBN 0-8057-8613-9 — ISBN 0-8057-8638-4 (pbk.)
    1. Evolution (Biology)   2. Natural selection.   I. Title.
    II. Series.
    QH366.2.B675   1993
    575.01'62'09 — dc20                                                93-431
                                                                       CIP

10 9 8 7 6 5 4 3 2 1 (hc)

10 9 8 7 6 5 4 3 2 1 (pb)

Printed and bound in the United States of America.

# Contents

# *Preface*

---

The study of the origins of Darwinism is an active area within the history of science. Many books and articles have been published over the past few decades, and more continue to pour from the presses. This level of interest is sustained by a conviction that the theory of evolution is not only one of the great achievements of science, but that it has also had major effects on the ways in which we think about ourselves, our society, and the world we live in. The topic is enmeshed in controversy, not merely because the historical issues are complex, but because they integrate directly into our feelings about modern life. No historian can claim to be objective when writing about a topic as controversial as Darwinism; the best one can do is to ensure that the reader is alert to the ways in which history has been manipulated by modern interests.

In the past twenty years some of the underlying interests that shape our views of Darwin and his theory have become clearer—even if divisions remain unresolved. Certain interpretations once taken for granted have been exposed as myths erected by scientists or the opponents of materialism. The current level of debate retains its emotional basis, but at least now we are all aware of what is going on. Yet for the general reader, this welter of controversy may seem merely confusing. Why bother to study a topic that seems incapable of resolution? Because to turn away from an important area of the history of science and culture simply because it is controversial is dangerous; it leaves old myths untouched so that they can continue

---

to be used to manipulate public opinion. Historians, and scholars generally, have a responsibility to ensure that their debates do not remain the exclusive domain of academics in their ivory towers.

In this survey of the history of Darwinism, I have tried to provide an outline of some of the major reinterpretations of the subject that have emerged in recent years. The fact that I myself have been involved in some of these developments means that my account cannot be impartial, but I have tried to show why the issues have been raised, so that readers will appreciate the need for vigilance in approaching any account of Darwinism, my own included.

# Acknowledgments

Chapter 4 of this book is a substantially modified version of the text of one of my lectures on "Biology and Social Thought, 1850–1914," given at the International Summer School in History of Science at Uppsala, Sweden, in June 1990. Copyright for the text of these lectures is held by the University of California Press for the Office of the History of Science and Technology at the University of California, Berkeley, and the material is used with permission.

*Portrait of Charles Darwin, 1854*

# Chronology

1809  Jean-Baptiste Lamarck's *Philosophie zoologique* presents a comprehensive theory of "transformism." Charles Darwin born.

1827  The radical Edinburgh Lamarckian Robert Edmund Grant moves to London and comes into conflict with Richard Owen over transformism.

1830  First volume of Charles Lyell's *Principles of Geology* interprets earth history as a process of gradual change; final collapse of the Biblical view of the earth's origin and Noah's flood.

1831  Darwin leaves England on HMS *Beagle*, embarking on a five-year voyage of discovery.

1844  Robert Chambers's *Vestiges of the Natural History of Creation* tries to make progressive transformism palatable to middle-class opinion.

1851  Herbert Spencer's "The Development Hypothesis" advocates Lamarckism; his *Social Statics* expounds the philosophy of progress via free enterprise.

1858  The Darwin and Wallace papers delivered to the Linnean Society.

1859  Publication of Darwin's *On the Origin of Species; or, The Preservation of Favoured Races in the Struggle for Life.*

1863  Thomas Henry Huxley's *Man's Place in Nature* stresses the similarities between humans and apes; Lyell's *Antiquity of*

*Man* popularizes the belief that the human race is much older than allowed by the Biblical time scale.

1864   Spencer's *Principles of Biology* integrates natural selection and Lamarckism into a comprehensive philosophy of biological and social progress.

1866   Ernst Haeckel's *Generelle Morphologie* advocates a radically materialist interpretation of progressive evolution.

1871   Publication of Darwin's *The Descent of Man and Selection in Relation to Sex.*

1879   Samuel Butler's *Evolution Old and New* attacks Darwin and proclaims Lamarckism a nonmaterialistic philosophy of evolution.

1882   Darwin dies and is buried in Westminster Abbey.

1893   August Weismann's *The Germ Plasm* stresses the role of "hard" heredity and precipitates a debate on Lamarckism.

1900   "Rediscovery" of Gregor Mendel's laws of heredity; Hugo De Vries goes on to advocate evolution by sudden "mutations."

1901   Karl Pearson's *National Life from the Standpoint of Science* argues for state control of human breeding to prevent degeneration; the eugenics movement gains popularity.

1911   William Johnson Sollas's *Ancient Hunters* depicts the "lower" races as relics of prehistory swept aside in a racial "struggle for existence."

1930   Ronald Aylmer Fisher's *Genetical Theory of Natural Selection* published.

1944   Julian Huxley's *Evolution: The Modern Synthesis* and George Gaylord Simpson's *Tempo and Mode in Evolution* consolidate the synthesis of Darwinism and genetics.

1975   Edward O. Wilson's *Sociobiology* precipitates a controversy over the use of natural selection to explain human behavior.

# 1

# From Darwin to Modern Darwinism: An Overview

Darwinism is a scientific theory, but it is also a theory that has had enormous impact on the way we think about ourselves. It was Charles Darwin who challenged the traditional view of divine creation and taught us that we are the products of a long series of purely natural developments within the animal kingdom. If scientists and materialists have welcomed this rejection of what they regard as outdated superstition, religious thinkers and other opponents of materialism have seen Darwin's theory as a threat. By reducing us to the status of animals, the opponents believe, Darwinism has encouraged us to behave like animals, overthrowing traditional moral values and turning society into a free-for-all struggle.

These contradictory attitudes emerged shortly after the publication of Darwin's *Origin of Species* in 1859, and they are equally strong today. The historian seeking to evaluate the origin and development of Darwinism cannot help but be affected by current prejudices. One of the major problems encountered in the study of Darwinism centers on the need to evaluate very cautiously the oversimplified historical myths generated by the strong feelings that Darwin's theory has aroused. But reaching an understanding of Darwinism requires more than a mere effort to see behind the caricatures erected by interested parties. Obviously there are contradictory myths involved, but less obviously, the source of the mythologies has changed significantly since the theory was first popularized.

Darwinism is not a simple entity, even within science. The term *Darwinism* can refer to the general idea of biological evolution, and it is also used more specifically to denote the mechanism of evolution proposed by Darwin, natural selection. One can believe in evolution, however, without believing that natural selection is the main or even a significant cause. Outside science, the label "Darwinism" has been applied to a variety of movements, especially in the social field. Various policies, usually of the political right, have been designated as forms of "social Darwinism." Since Darwin's name sprang to prominence in the 1860s, it has been applied to a series of different positions that have, not surprisingly, evolved in the course of time. In some areas, the meaning of "Darwinism" has changed almost beyond recognition. The "Darwinism" of the 1860s is not the "Darwinism" of the modern biologist, nor of the modern sociologist. The historian must constantly beware the imposition of inappropriate labels on the past and must realize that developments within science and the public debate have changed the context within which Darwinism is evaluated.

## Darwin and Natural Selection

To most modern biologists Darwin is a hero of discovery, because he pioneered both the general theory of evolution and the most successful explanation of how evolution works. The publication of the *Origin of Species* certainly precipitated the revolution by which the development of life on earth came to be regarded as a topic of scientific study rather than a theological question. The *Origin* also introduced natural selection, which Darwin regarded as the chief mechanism of evolution. Assuming random variation among the individuals making up a population, Darwin argued that some characteristics will, by chance, be better adapted to changes in the environment than others. Given the "struggle for existence" caused by the tendency toward overpopulation, the better-adapted individuals will survive and breed, passing their advantages on to their descendants, who will make up an ever-greater proportion of succeeding generations. Over a long period of time this natural selection of random variation will produce a significant change in a population, enough to turn it into a new species.

Following the rediscovery of Gregor Mendel's laws of heredity in 1900, it was shown that genetic mutation and inheritance explain the random variability assumed by Darwin, and the transmission of advantageous traits to later generations. The "modern synthesis" of Darwinism and genetics was established by the 1940's and has remained the dominant theory of the evolutionary process. This belated success of the selection theory has, however, created problems for the historical interpretation of Darwinism. Modern biologists are tempted to assume that since Darwin proposed the correct explanation of the evolutionary mechanism, his success in this area must explain why he was able to convince his contemporaries that the general idea of evolution was acceptable. They believe that the discovery and publication of the selection theory marked the turning point that converted the general idea of evolution from a mere speculation to a scientific hypothesis, and in this view Darwin's initiative becomes the crucial step forward in the history of evolutionary biology.[1]

Modern historical work, however, has begun to cast doubts on this interpretation. It has become increasingly obvious that Darwin succeeded in converting the world to evolutionism not *because* he had the theory of natural selection, but *despite* the fact that most of his fellow biologists had major reservations about it. Natural selection was certainly taken seriously by a few of Darwin's close supporters, but some of the biologists who played a key role in popularizing evolutionism were convinced that selection was not the primary mechanism of change. Even T. H. Huxley, who became known as "Darwin's bulldog," must be included in this category. The first generation of "Darwinists" saw themselves as followers of Darwin on the general topic of evolution, but not on the question of natural selection. In the later decades of the nineteenth century, many biologists became outspoken opponents of the selection theory, erecting a variety of alternative mechanisms of evolution.[2]

## Lamarckism

The most important alternative to natural selection was the mechanism of the "inheritance of acquired characteristics," often called

"Lamarckism" because it was advocated by the pre-Darwinian evolutionist J. B. Lamarck. In any discussion of the impact of Darwinism the distinction between natural selection and Lamarckism is vital—and in some circumstances the two are easily confused. Natural selection is based on the assumption that individuals are born with inherent differences, which we now explain in terms of genetic characters. Selection merely picks out those inherited factors that are by chance useful in dealing with the environment. Lamarck pointed out that animals are not passive in their interactions with their surroundings; if the environment changes they will adopt new habits in an attempt to cope with the changes, and in so doing they may acquire characteristics that they were not born with. Thus, to use the classic example, the ancestors of the giraffe may have encountered conditions in which their best source of food was the leaves of trees. In these circumstances they would develop the habit of stretching up to reach the leaves, and in the course of time their necks would have become somewhat longer—just as a weight lifter's arms become stronger due to exercise. Lamarck assumed that such acquired characters could be inherited: the next generation of giraffes would be born with slightly longer necks, and would of course continue the process. Over many generations the results of the animals' efforts would accumulate to give the species a new characteristic: the long neck of the modern giraffe.

According to modern genetics, acquired characters cannot be inherited, and so Lamarckism is not a valid mechanism of evolution. But no one of Darwin's generation knew this; even Darwin himself accepted that Lamarckism would supplement the actions of natural selection. Some of Darwin's supporters, including the philosopher Herbert Spencer, thought that Lamarckism was even more important than natural selection. Only in the last decades of the nineteenth century did there appear a small body of biologists who insisted that the Lamarckian effect did not work. They argued that selection was the only mechanism of evolution and became known as "neo-Darwinists." This was a very controversial position, however, and at the time, the majority of biologists thought that selection was at best only part of the story. Julian Huxley—T. H. Huxley's grandson and a founder of the Modern Synthesis—referred to the period around 1900 as the "eclipse of Darwinism," because of the strong antiselectionist feeling among biologists. Even genetics was at first seen as incompatible with natural selection, although it eventually eliminated Lamarckism and thus paved the way for the emergence of modern Darwinism.

Recognizing the initial failure of the selection theory forces the historian of science to ask new kinds of questions about the advent of evolutionism. We can no longer believe that Darwin popularized the general theory of evolution because his proposed mechanism was so successful. We are forced to recognize that the scientific community was persuaded to accept evolution even though it was widely believed that an adequate explanation of the process was lacking. This suggests that the motivations of those who participated in the scientific revolution were more complex than have been traditionally imagined. Evolutionism served the purposes of a new generation of scientists determined to wrest control of an important area of discourse away from the theologians. The Darwinism revolution owes as much to changes within the newly professionalized scientific community as it does to the persuasive effects of Darwin's arguments and evidence.

From this new historical perspective, the subsequent emergence of Mendelian genetics and its somewhat belated synthesis with the selection theory take on a new significance. Genetics did not merely "fill in the details" of a general program outlined by Darwin. Instead, there were later developments in evolution theory—developments just as fundamental as the original popularization of the theory by the *Origin of Species*. The elimination of Lamarckism was symptomatic of a major transformation in scientists' thinking on the questions of heredity, variation, and evolution.[3] If we wish to survey the history of Darwinism and its cultural implications, we must keep these scientific developments firmly in mind at all times.

## Religion

When the *Origin of Species* first appeared there was, understandably, a certain amount of opposition to the basic concept of evolution from religious thinkers who wished to preserve the idea of divine creation. Yet this kind of blanket rejection of evolutionism was comparatively shortlived. When Darwin died in 1882, he was buried in Westminster Abbey as a pillar of the Victorian establishment. The creationist opposition to evolutionism that is very active today, especially in the United States, is a twentieth-century phenomenon sparked by scientists' efforts to introduce evolutionism in the high-school curriculum. In Darwin's own time, what we would today call "creationism" was in retreat, as the majority of religious thinkers rushed to make an accommodation with evolutionism. Here again, we must be careful not to allow modern preconceptions to distort our interpretation of the past. Once we recognize that the Darwinism of the 1860s and 1870s represented merely generic support for evolutionism, we can appreciate that religious thinkers were in a position to accept it and to argue that evolution might in fact be God's own method of creation.

The crucial factor that made this kind of compromise possible was the belief that evolution was an essentially purposeful process. Evolution was understood to be inherently progressive, representing the steady ascent up a ladder or hierarchy of organic complexity leading toward humankind. The human mind and human moral values were seen as the intended outcome of a process that was built into the very fabric of nature and that could thus be interpreted as the Creator's plan. The first generation of evolutionists followed the philosopher Herbert Spencer in his belief that the history of life on earth represented the unfolding of a developmental sequence aimed at the production of higher things. Indeed, it was probably essential for the success of the theory of evolution that early evolutionists adhered to such a philosophy. Darwin gained respectability because his theory was interpreted as the scientific foundation for a more general view of progress in which divine purpose was expressed not in a supernatural act of creation, but in the creative activity of the laws of nature instituted by God.

The claim that Darwinism has always been inherently opposed to religion must thus be treated with great care. And yet there is a more restricted sense in which the hostility of many religious thinkers is justified. The theory of natural selection is much less easily squared with a view of evolution as the unfolding of a divine purpose. The "survival of the fittest" in a never-ending "struggle for existence" sounds more like a game of genetic Russian roulette than the kind of mechanism that would have been instituted by a wise and benevolent Creator. Historians are still arguing over whether Darwin himself realized (as his modern followers claim to) that natural selection is not inherently progressive and offers no predictable goal toward which evolution is aimed. He may well have recognized this point as a biologist, but when he addressed the human implications of his theory Darwin tended to slip into progressionist language. His followers popularized evolution-ism by very carefully minimizing the role and the wider implica-tions of natural selection.

From the start, however, there were some religious thinkers who saw that natural selection made nonsense out of any claim that na-ture was the expression of a divine purpose. Even some of Darwin's supporters, including the American botanist Asa Gray, thought that God might steer evolution in the right direction by somehow influencing the direction of variation within the species. From a sci-entific perspective, this possibility was little better than creation-ism, and it was soon replaced by naturalistic theories that seemed more compatible with divine benevolence. The most important of these alternatives, Lamarckism, reintroduced purpose into evolu-tion by allowing that animals direct the course of their species' fu-ture development through their choice of new habits. An element of Lamarckism was incorporated into what passed for "Darwinism" in the 1860s, but in later decades "neo-Lamackism" emerged as a ri-val to "neo-Darwinism." Many opponents of materialism, some of whom had no direct religious axe to grind, took up the inheritance of acquired characteristics as a more humane and more purposeful form of evolutionism. The novelist Samuel Butler was one of the first to advocate this position as part of a direct attack on the philos-ophy of neo-Darwinism.

Although Lamarckism has been almost banished from science to-day, its supporters were effective in shaping public perceptions of the rival theories. There is a widespread belief that Darwinism (the selection theory) represents an inherently materialistic philosophy in which nature is reduced to a meaningless sequence of random events. Moral values have no role in such a philosophy; mere sur-vival and the enjoyment of physical pleasures become the only pur-poses in life. Anti-Darwinism theories such as Lamarckism are presented as the basis for an alternative philosophy of life that re-tains the purposefulness of nature and endorses traditional moral values. From the historians' perspective, however, these attempts to pin ideological labels onto the theories are very misleading. It is a historical fact that Lamarckism has been used to endorse all the un-pleasant moral consequences normally associated with Darwinism, including support for racism, imperialism, and unrestrained free-enterprise. By recognizing that anti-Darwinian theories have played a role in the historical development of evolutionism, we are forced to confront the possibility that many of the consequences that have been attributed to Darwinism should in fact be distributed around a much wider range of evolutionary philosophies.

## Social Darwinism

This is particularly true in the sphere of social thought, where anti-Darwinian sentiment has created the image of a ruthless "social Darwinism" dominating late nineteenth-century thought. It is claimed that, by portraying the human race as a product of natural evolution, Darwin left us with no alternative but to treat success as the only measure of what is "good" in nature's terms. The image of a revived "social Darwinism" is still used today by writers who wish to highlight what they see as the unpleasant aspects of certain theories.

When Edward O. Wilson began to apply the techniques of sociobiology to human nature in the 1970s, his efforts were attacked for being merely a new form of "social Darwinism."[4] In this case, the term is literally correct: sociobiology seeks to explain the behavior of social animals in terms of instincts implanted by natural selection. If it could be shown that human beings are influenced by such instincts, then the Darwinists would indeed have something to say about social policy. Wilson's views were violently opposed by political and social thinkers who insisted that human behavior is not constrained by inherited instincts. They pointed out that the instincts he postulated conveniently seemed to imply that a society based on individual competition was natural and hence inevitable. On the surface Wilson seemed to be reviving the free-enterprise form of social Darwinism pioneered by Herbert Spencer in the nineteenth century. But to depict Wilson's views as merely a continuation of a long-standing tradition is to miss the fact that sociobiology is one of the first genuinely Darwinian—in the strict, modern sense of the term—applications of the theory of natural selection to human affairs.

Because non-Darwinian theories were more popular than Darwinism itself in the late nineteenth and early twentieth centuries, we ought to pause before we uncritically accept the assumption that all applications of biology to social policy were based on the selection theory. Equally problematic is the wide range of policies that have attracted the name "social Darwinism." The term has been applied to a number of political philosophies that, in one way or another, advocate a role for struggle or competition in society. It is easy to see why this connection is made, given that Darwinism is associated with phrases such as "struggle for existence" and "survival of the fittest." But struggle can be imagined to occur between individuals within a population (as in unrestrained free-enterprise) or between distinct populations (conflict between races or nations). Darwinism could certainly be used as the basis for arguing that struggle at each of these levels is necessary for progress—but so could most of the rival theories.

Whether the debate over its influence is conducted at the level of morality or social theory, Darwinisn has been seen as the principal culprit in the process by which biology has been misused to support political—usually right-wing—policies. This impression has been so strong that "social Darwinism" has become a favorite term of abuse among both left-wing and liberal thinkers. Any effort by the historian to point out that the situation is rather more complex is likely to be greeted with the charge that he or she is trying to set up an unnaturally sharp dividing line between science and social values. It is widely assumed by the critics of Darwinism that the theory of natural selection is itself a reflection of Victorian social values. From Marx and Engels onward, it has been fashionable to see the selection theory as an inevitable product of nineteenth-century capitalism. Only within a society dominated by the image of individual competition could such a model of natural processes have been created. To challenge the validity of "social Darwinism" is to support the naive views of scientists who imagine that new theories are created purely to explain the facts of nature.

This is not the place for a discussion of the general question of the objectivity of science.[5] But in the particular area of evolution theory it is difficult to escape the feeling that theories have reflected social values, if only because the images and metaphors used to describe the hypothetical natural processes are so often drawn from human life. Not surprisingly, these metaphors are then used to vindicate social policies by means of the claim that such policies are "only natural." "Darwinism" both reflected and was used to endorse the values of a free-enterprise society. But there were other evolutionary processes under discussion at the time, and it would be surprising if they were somehow exempt from the same interaction with social values. A critique of "social Darwinism" must be based not on a desire to defend the objectivity of science, but on a recognition of the fact that far more than Darwinism was involved. In its original form, "Darwinism" was a synthesis that included many processes that would not be accepted by a modern biologist. The subsequent history of the interaction between biology and social thought reveals a complex network of relationships between many different scientific theories and an equally diverse range of social policies.

The widespread belief that there is a unitary Darwinian theory that promotes the use of the metaphor of the "survival of the fittest" in social thought thus turns out to be a myth. Darwinism has been too convenient a target; it has been blamed for a wide range of influences, some of which have very different scientific and cultural roots. Those who have wished to highlight the ruthlessness of social policies to which they object have used the term "social Darwinism" in the expectation that this will conjure up an image of an amoral philosophy based on the worship of struggle and success. Without fully vindicating Darwinism, it must be shown that there are other theories that should also be taken into account. Much of what has passed for "social Darwinism" has only the loosest connection with what the modern biologist regards as Darwin's most important scientific insight, and in some cases it reflects theories that were positively hostile to Darwin's. Mere use of the concept of struggle does not qualify one as Darwinian if the same concept was used by Lamarckians and other exponents of non-Darwinian evolutionary processes.

## Darwinism and Capitalism

The classic form of social Darwinism is certainly the link with free-enterprise individualism. Nineteenth-century advocates of laissez-faire economic policy, such as Herbert Spencer, believed that governments should impose no restrictions on the activities of the individuals who make up society—each should be left to pursue his or her own self-interest, and the "natural" interactions that would result would be the ones most likely to promote economic and social progress. It is easy to see why the theory of natural selection would be seen as a projection of this social philosophy onto nature. In Darwin's theory, individual animals struggle against one another to determine which will live and which will die, and the success of the most able individuals forms the basis for evolutionary change within the population. Natural selection certainly does seem to reflect the individualist model of society, and there can be little doubt that Darwinism was at first linked to Spencer's extremist, free-enterprise social philosophy. Social progress was seen as a natural continuation of biological evolution, based on the same mechanism of change.

A complication arises, however, in that Spencer himself was as much a Lamarckian as a Darwinian in his biological evolutionism. How can the archetypal "social Darwinist" have been a Lamarckian? The answer to this question lies in the synthetic character of early Darwinism and in the complex nature of the concept of the struggle for existence. In Spencer's view struggle was not just a means of eliminating the congenitally unfit; it was also a means of stimulating every individual in the population to maximum effort. The penalty for failure was not necessarily death; more often the individual survived and learned from experience how to do better next time. Struggle was the spur to self-improvement, and the accumulation of individual acts of self-development led to social progress. Thus the concept of the struggle for existence was linked to the Lamarckian mechanism of the inheritance of acquired, or learned, characteristics. Spencer's "social Darwinism" was really a form of "social Lamarckism"—but because everyone associates the struggle metaphor with Darwinism, this has gone largely unnoticed. Nor did Spencer's philosophy value success at any price: he was convinced that struggle would promote the traditional Protestant virtues of thrift and initiative, thus ensuring moral as well as economic progress.

## Darwinism and Race

Lamarckism also played a major role in the thinking of those biologists and anthropologists who erected a hierarchy of human races based on the assumption that the white race was the pinnacle of evolutionary progress. Darwinians were not the only ones who believed that black peoples were relics of prehistory who had survived into the present only because they had been protected by geographical isolation. But surely Darwinism must have helped to sustain the assumption that when different races come into conflict they must struggle to determine which should occupy the land? It is certainly true that Darwin and his followers invoked the image of successful new species spreading out from their original home and exterminating the less evolved forms in the surrounding territory. Late nineteenth-century explanations of the geographical distribution of species were pervaded with the metaphors of imperialism: conquest, colonization, and extermination. But the same metaphors were also used indiscriminately by anthropologists and biologists who rejected Darwinian natural selection as a mechanism of positive, creative evolution. To many thinkers at the time, selection worked only as a negative process: it could eliminate the less successful of nature's experiments, but it did not explain how new species actually came into existence.

When we look at the cultural background of those who invoked the struggle for existence at the level of nations and races, we find that it often reflected a conservative tradition diametrically opposed to the individualism that is seen as the cultural foundation stone of Darwinism. To understand this, one must remember that the exponents of nationalism and imperialism often rejected the ethics of the free-enterprise system, which had originally been based on the expectation that industrialization would eliminate national and racial differences. There is a school of historical thought which believes that each period in the development of the human mind has been dominated by a particular race or culture which was, at that time, the most creative. This view of the past played a major role in nineteenth-century thought, in opposition to the theory of continuous, gradual development that is usually seen as the cultural matrix of Darwinism.[6] And this rival historiography could just as easily generate a social philosophy based on racial or national struggle.

A group of very different social philosophies has thus been linked together merely because each advocated a role for struggle in human affairs. They have been artificially unified under the name "social Darwinism" by writers who object to all of them, and who have nothing to gain from admitting that the debate is actually a good deal more complex that their simple polarization would imply. Those who object to unrestrained free-enterprise, racism, and nationalism have found Darwin's name a convenient label with which to characterize all the positions they find so repulsive. If scientists have exaggerated Darwin's role out of respect for the modern theory of natural selection, these opponents of "Darwinism" have made the same mistake by erroneously attributing a wide range of moral and political philosophies to the influence of his theory alone.

This study of Darwinism will thus serve not as a means of unifying the history of Western culture but of proclaiming its diversity. The nineteenth and early twentieth centuries were not dominated by a single social philosophy, nor was the science of the time dominated by a single theory of evolution. The emergence of evolution theory in biology, and the widespread use of evolutionary metaphors in social thought, reflect complex developments within a culture that was just as fragmented as our own. The search for simple unifying factors, for symbols that will somehow capture the essence of a historical period, is illusory. Societies change from decade to decade as the balance of power shifts among antagonistic social and cultural forces, and as new forces come on the scene. Looking into the past may well help to throw light on the origins of modern problems and attitudes, but only if we take into account the fact that our current situation is the outcome of genuine historical development. Simply projecting modern prejudices back onto the past will obscure rather than clarify the issues. History must become a tool for seeing through the myths and preconceptions erected by those who wish to use oversimplified images of the past as a means of conducting modern debates.

There can be little doubt that evolution theory has had an enormous impact on the way we think about human nature and human behavior. Evolutionism made possible entirely new interpretations of the origin and nature of human moral values and society. But it was not simply an invasion of a traditional, religious culture by rampant materialism. We cannot treat the late nineteenth century as a period that was bullied by a single scientific theory into accepting a philosophy of moral nihilism. A whole range of social and cultural positions adjusted to the new situation and made use of the new metaphors of evolutionary development. Conservative, liberal, and radical thinkers, each in their own way, took up the notion of progress through struggle. Only by acknowledging the complexity and the ever-changing character of Darwinism shall be we able to make sense of its impact on the modern world.

# 2

## Darwinism in Science

Darwin's name dominates the history of evolution theory in a way that makes it difficult to outline a broadly based approach to the subject. Yet it is now becoming obvious to historians of science that the emergence of modern evolutionism involved factors that are scarcely mentioned in conventional accounts of the "Darwinian revolution." The attempt to create a less Darwin-centered view of the development of evolutionism is not intended to discredit Darwin himself. Indeed, it might be argued that in some ways Darwin was "ahead of his time"—his conception of the evolutionary process was not fully absorbed by his contemporaries and had to wait until the mid-twentieth century for its full expression in biology. But once we accept that some of Darwin's most important insights were not taken up in his own time, we can begin to ask more sophisticated questions about the role played by the *Origin of Species* in precipitating what is often called the "Darwinian revolution." The biologists of the 1860s and 1870s did not accept evolutionism in the same sense and for the same reasons that biologists now accept modern Darwinism. The questions we ask about this scientific revolution must take into account the very different context within which the theory was first evaluated.

There are certain popular misconceptions that are fairly easily disposed of by outlining the state of science in the period leading up to the introduction of Darwinism. It is often assumed that Darwin had to confront a culture that still accepted a literal interpretation of the story of creation found in Genesis. In fact the geologists and paleontologists of the early nineteenth century had demonstrated that the earth was immensely ancient and had undergone a vast sequence of changes leading to its modern state. Few educated people in 1859, when the *Origin of Species* was published, still believed that the world had been created only a few thousand years before. It was generally accepted that life had developed through time, leading up to the earth's modern inhabitants; what was in question was how the process of development had taken place.[1]

Less easily dispelled are the myths that have grown up surrounding the debate in which scientists were persuaded to accept that the origin of new species was a natural rather than a supernatural process. Here the English-language literature is dominated by an orthodox picture of the rise of evolutionism that is only slowly yielding to criticism based on a more sophisticated level of analysis. It is not that the conventional picture is wrong, but it is incomplete in a way that significantly distorts our understanding of the events and their significance. Perhaps the best way of characterizing the traditional picture is that it is a view of the history of evolutionism which emphasizes the importance of the *mechanism* of organic change at the expense of a wide range of other evolutionary topics. The centerpiece of the history of evolution is the theory of natural selection, because this hypothesis for how evolution works is seen as Darwin's greatest contribution (by the modern biologists) and his most culturally disruptive idea (by its opponents).

The orthodox historiography concentrates on the events leading to the discovery and publication of the theory of natural selection, highlighting those discoveries by Darwin and his predecessors that helped him to formulate this radical new proposal about nature's operations. The story of the "Darwinian revolution" is told in terms of the debate that followed the publication of the *Origin of Species*, especially the controversy about the mechanism of change. Once that debate was settled in favor of the Darwinians, attention shifts to later developments that filled in the missing pieces of the jigsaw puzzle which Darwin himself had so nearly completed. Since we know that Darwin did not anticipate Mendel's theory of particulate inheritance (unit characters produced by what we now call the genes), the story concentrates on the developments in heredity theory that paved the way for the "rediscovery" of Mendel's laws, on the emergence of genetics in the early twentieth century, and on the subsequent amalgamation of genetics and the selection theory that gave rise to the Modern Synthesis, which still dominates biologists' thinking.

What is missing from the orthodox picture is a series of events that, when taken into account, force us to adopt a very different interpretation of Darwin's role. It is now clear that the basic idea of evolution had been widely debated before the *Origin of Species* was published. Darwin introduced his theory into a culture that was very well aware of the religious, moral, and political implications of evolutionism. Many biologists had already begun to realize that creationism was a theological device for excluding them from this area of discourse, and as professional scientists they were anxious to proclaim their new role in society by invading a territory from which they had for so long been barred. Because many of the scientists came from the newly powerful middle class, they supported efforts to present history—including the history of life on earth—as a steady progress toward higher levels of development. The industrial progress that was giving the new commercial and professional classes an ever greater role in society could, in turn, be presented as the continuation of a fundamental trend in nature. It is only when these social factors are taken into account that we can explain why Darwin's theory became the figurehead for the new evolutionism, even though most of his fellow biologists found natural selection unacceptable.

Within the scientific debate itself, we are now beginning to realize the limitations of many traditional accounts. Natural selection was not, at first, the centerpiece of "Darwinism." Darwin's hypothetical mechanism served as the catalyst that converted scientists to evolutionism, but it was too controversial to satisfy them. The problem of heredity, so central to the selection-dominated model of the debate, was at first of only limited interest. A small handful of naturalists debated with Darwin the viability of his selection theory, but most accepted the basic idea of evolutionism without worrying very much about the mechanism of change.

Some of the most active areas of evolutionary biology are scarcely mentioned in conventional accounts of the Darwinian revolution. A whole generation of biologists used morphology (the study of animal structure through comparative anatomy and embryology) in an attempt to reconstruct the relationships between the various branches of the evolutionary tree. Paleontologists also reconstructed the origin and history of various groups from the increasingly large number of fossils that were being discovered—although their work has been largely ignored, apart from a few debates over "missing links." Biogeography, the attempt to trace the point of origin and the subsequent migration of the various groups of plants and animals, became an important research field in the last decades of the nineteenth century.

These branches of evolutionary science have been largely ignored by historians because they do not fit into the conventional focus on natural selection. Most morphologists and paleontologists either did not participate in the debates over the adequacy of natural selection or opted openly for non-Darwinian mechanisms such as Lamarckism. It is precisely because these areas of science have gone largely unrecorded that it has taken so long for us to appreciate the extent to which late nineteenth- and early twentieth-century biology was dominated by non-Darwinian views of evolution. Yet these areas supplied some of the most important metaphors that would be used to justify the moral and social applications of evolution theory. Morphologists and paleontologists talked about evolutionary progress and degeneration. Biogeographers visualized newly evolved forms spreading out and conquering the surrounding territory, driving the previous inhabitants to extinction. Although some of these metaphors were used by Darwinians, they were also used by biologists who had little interest in the selection theory.

We certainly should not ignore the debates over natural selection and heredity, but neither should we be blinded by an overconcentration on those areas and neglect the other branches of evolutionary science that helped to determine how Darwin's contemporaries responded to the general idea of evolution. When modern genetics and the modern Darwinian synthesis emerged in the twentieth century, they did so among scientists who were consciously reacting against the first generation of evolutionary biology. To understand the response of Western culture to Darwinism, we need to have some familiarity with both the conventional history of natural selection and genetics and the less well-known story of the various other attempts that were made to reconstruct the history of life on earth.

## Evolution before Darwin

In the English-speaking world, at least, early nineteenth-century views on the origin of life were dominated by the "argument from design." As expressed in William Paley's classic of 1802, *Natural Theology,* this argument attempted to demonstrate the existence of a wise and benevolent Creator by showing that the various animal forms must have been designed by a supernatural intelligence. Just as the complexity of a watch proved the existence of a watchmaker, it was thought that the complexity of animal and human bodies proved the existence of a supernatural designer, because no purely natural process could generate such perfectly functioning structures. The fact that each species is well adapted to its environment also proved that the Creator was benevolent; obviously, He intended His creatures to live a happy life.

Paley's rather simple version of the argument from design was no longer very popular with French and German naturalists, although the conservative thinkers among them assumed that something more than a purely natural process was required to create new forms of life. In any case Paley's vision of a perfectly adapted creation was breaking down in the face of geologists' demonstration that the earth had changed considerably over a vast period of time. The fossil record also indicated that there had been a whole series of animal and plant creations before the present one. Various efforts were made to preserve at least some elements of the traditional world view in the face of this new evidence. Many geologists thought that the earth's history had been punctuated by vast catastrophes that had raised new mountain ranges almost instantaneously. It was thought that such great upheavals would have wiped out all living things, possibly over the whole of the earth's surface, thus requiring a sequence of divine creations to repopulate the earth.

As a geological theory, this "catastrophism" was by no means as backwards as historians of science once imagined, but it was certainly incompatible with a theory of gradual evolution in the organic world. It was challenged by the Scottish geologist Charles Lyell in his *Principles of Geology,* which appeared in three volumes between 1830 and 1833. The subtitle of Lyell's book outlines his alternative view of geological change: *An Attempt to Explain the Former Changes of the Earth's Surface, by Reference to Causes Now in Operation.*[2] He insisted that there was no need to invoke catastrophes; the observable processes of earthquakes, erosion by streams, and the like, could explain all the complex geological formations, provided they were supposed to have operated over vast periods of time. Lyell's "uniformitarian" geology depicted the earth as a system that had not changed significantly over time—it had always operated in pretty much the same way. Individual mountains and continents were gradually being destroyed, and others were being formed, equally gradually, to replace them.

Lyell's book had an immense influence on Darwin because it challenged the catastrophist theory of sudden extinctions followed by mysterious recreations of life. Lyell himself did not accept what we would now call evolution because he could not face the prospect of linking the human race to the apes. But anyone who, like Darwin, converted to Lyell's unformitarianism was forced to confront the possibility that change in the organic world might be equally gradual and natural. To demonstrate his own rejection of this suggestion, Lyell attacked the theory of "transformism" (evolution) proposed by the French biologist Jean-Baptiste Lamarck in books such as his *Philosophie zoologique* of 1809.

Lamarck had provided one of the first complete philosophies of natural development in the organic world. He supposed that the simplest forms of life were spontaneously generated from nonliving matter. Then, over many generations, life gradually mounted the scale of complexity until at last the human race was formed. To explain how living forms became adapted to changes in their environment, Lamarck proposed the mechanism of the inheritance of acquired characteristics (see in chapter 1). In his theory, adaptation was a natural process, not a fixed state instituted by the Creator.[3]

Historians used to believe that Lamarck was almost completely discredited by his rival, Georges Cuvier. Cuvier was an anatomist who pioneered the study of fossil remains and also helped to found the catastrophist theory of geology. It is now recognized that Lamarck's theory was actively debated in France throughout the early decades of the nineteenth century. In Great Britain, the radical Edinburgh anatomist Robert Edmond Grant supported Lamarck in his attempt to discredit the orthodox view of living nature. When Grant moved to London in 1827, he continued to argue for transformism as a means of urging the reform of medical teaching. He was also known as a political radical who wished to overthrow the existing social hierarchy. Richard Owen and other conservative naturalists organized a lengthy, and ultimately successful, resistance to Grant's attack. Lamarckism was portrayed as an integral part of materialist philosophy, dangerous because it threatened the moral and political order of society. Darwin had met Grant during his brief stay in Edinburgh as a medical student. He must have been only too well aware of Grant's eventual disgrace and of the dangerous implications that had come to be associated with transformism.

In 1844 the Edinburgh publisher Robert Chambers made a new attempt to force middle-class opinion to take transformism seriously. His anonymously published *Vestiges of the Natural History of Creation* tried to uncouple the theory from materialism by claiming that the ascent of life toward mankind could be seen as the unfolding of a predetermined scheme of development programmed into nature by its Creator. Borrowing an argument from Charles Babbage, the inventor of a mechanical forerunner of the computer, Chambers thought that God had designed the universe to unfold in a predetermined direction. The emphasis on progress was important because it gave transformism the air of purpose that would make it acceptable to people who saw themselves as participants in the progressive development of industrial society. The scientific details of Chambers's book were crude, and it was harshly condemned by naturalists committed to Paley's static argument from design. But the implications of the theory were now firmly on the agenda for debate as Britain moved toward an age of prosperity and industrial expansion.

The biologists who opposed Chambers's book stressed that the fossil record did not support transformism. There was, it had to be admitted, a progress from the simplest to the highest forms of life in the course of the earth's history. But it was not a continuous progress: there were great steps in the ascent of life that could not be explained by gradual transformation. Outside of these steps, however, even conservative naturalists were beginning to accept that there were patterns of orderly development in the sequences of fossil species. Richard Owen himself argued for a process of branching or divergence in which the members of each class gradually moved out into a series of different adaptive specializations. To Owen, such patterns were part of a divine plan of creation; to Darwin they represented evidence for a process of adaptive evolution. In the course of the 1850s, naturalists came closer to acknowledging that there were regularities in the course of the history of life, although they were (quite rightly) suspicious of the simple ladder of progress advocated by the Lamarckians and by Chambers.

When the *Origin of Species* was published in 1859 it did not come as a bolt from the blue. On the contrary, Darwin carefully orchestrated the presentation of his theory so that it would fit in with a growing trend toward acceptance of a more naturalistic view of the history of life. What was new in the *Origin* was the mechanism of natural selection, and we are now in a position to see why this might have been decisive, even though it was not widely accepted. In 1851 the social philosopher Herbert Spencer had published an article on "The Development Hypothesis" in which he supported the Lamarckian mechanism. Spencer believed that progress was inevitable within a free-enterprise society, and he saw how biological evolution could be made to underpin a more generally progressive view of history. But Lamarckism had been too firmly discredited for biologists to take Spencer seriously, and Chambers's 1844 book had presented no new ideas for the mechanism of change. There was thus a crucial gap in the argument for evolution: before scientists could accept it, they needed to be offered a new suggestion as to how it worked. Darwin provided such an initiative, and although natural selection did not prove acceptable in the short run, it was enough to break the logjam and let the current of support for transformism flow on to dominate scientific thought.

## The Darwinian Revolution

Darwin had been converted to evolutionism by the evidence gathered on a five-year voyage of discovery as the naturalist on board a Royal Navy survey vessel, HMS *Beagle*. The *Beagle* spent much of her voyage (1831–36) charting the waters of South America, while Darwin traveled inland, studying geology and natural history. There he discovered evidence favoring Lyell's uniformitarianism, and on his return to England he made important contributions to the geological debate. The *Beagle* also called in at the Galápagos Islands, several hundred miles west of South America, in the Pacific. Historians have now questioned the old idea that Darwin experienced a sudden moment of discovery on the Galápagos Islands, but it is clear that as he reflected on his observations he realized that simple creationism was no longer plausible.[4] In the case of the birds he observed, most notably the mockingbirds and the finches, each of the relatively small islands was occupied by a slightly different species. To Darwin it was inconceivable that the Creator had performed a separate miracle on each island. He became convinced that the species had descended from small populations of South American birds that had been accidentally blown across the ocean. They were related because each shared a common ancestry in the parent South American form, but different because each had evolved in isolation on its own island.

Like Lamarck, Darwin saw that adaptation must be a natural process, not a divinely created state. But he was suspicious of Lamarck's suggested mechanism, not least because it was now regarded with such evident disfavor by the respectable scientists he most wanted to impress. Darwin set out to look for an alternative, and in the course of his researches began to study the activity of animal breeders. Historians are still arguing over the exact process of discovery, but eventually Darwin realized that the breeders were able to make significant changes in their herds and flocks by selecting and breeding from individuals who possessed a characteristic they wished to enhance. Over many generations, the characteristic could be developed by the process of artificial selection until eventually a distinct variety was formed.

Darwin then realized that a natural equivalent of this selective process could be operating in the wild, brought about by the intense competition for food and other resources. The social philosopher Thomas Malthus had argued in *An Essay on the Principle of Population* that the human species always tended to breed faster than the food supply could expand. Darwin realized that animals have the same potential for population expansion, yet their food supply is fixed by the environment. The result, he decided, must be a constant "struggle for existence" in which many perish in every generation. If there is any variation of character among the individuals making up the population, those best adapted to changes in the environment would be most likely to survive and breed. This process of "natural selection" would gradually intensify any character that better adapts the population to its new environment. Given time, Darwin believed, this process would produce a new species: a population differing so markedly from the parent form that interbreeding would no longer be possible even if geographical barriers were removed.

By 1842 Darwin was ready to write out a sketch of his new theory, and two years later he prepared a substantial essay.[5] But he was unwilling to publish it, partly because he knew that his ideas still needed further refinement, and partly because the controversy over Chambers's *Vestiges* had highlighted the divisive nature of evolution theory. Instead, he began to correspond with an ever-widening network of naturalists who could provide him with the information he needed to test his theory. To a few selected confidants, including the botanist Joseph Dalton Hooker, the details of the theory itself were slowly revealed for analysis and criticism. By the mid-1850s Darwin was at last persuaded that the time was ripe for publication, and he began to write what would have been a multivolume account of his theory.[6]

This "big species book" was never completed, however, because the writing was interrupted in 1858 by the arrival of a paper from another naturalist, Alfred Russel Wallace, who was on a collecting trip in the Far East. Wallace had independently thought up a mechanism to explain evolution, and sent an account of it to Darwin—who was widely known to be working on the problem. The fact that another naturalist should have independently hit upon the idea of natural selection is often cited by historians as evidence that the time was ripe for the idea. But it must be remembered that Darwin had been working on his theory for 20 years, during which time both the scientific and the cultural environment had changed significantly. A close reading of Wallace's paper also suggests that there were significant differences between his conception of natural selection and Darwin's mature theory. Whatever the differences, however, Darwin was convinced that he had been forestalled, and he asked his friends Lyell and Hooker what to do. At their suggestion, papers by Darwin and Wallace were read to the Linnean Society in London that year; and Darwin began to prepare the single-volume account of his theory that we know as the *Origin of Species*.

The argument of the *Origin* will be discussed in more detail below (chapter 5); here we must concentrate on the book's impact upon the scientific community. Thanks in part to Darwin's prepublication circulation of his ideas, a small group of biologists sprang to defend the theory against anticipated attacks. Hooker played an important role, along with the young and ambitious Thomas Henry Huxley, who had just put his foot on the first rung of his career as a professional scientist (he taught paleontology at the Royal School of Mines in London). Huxley would eventually become known as "Darwin's bulldog," and it was he who featured most prominently in one of the most famous confrontations of the debate when, at the 1860 meeting of the British Association for the Advancement of Science, Darwin's theory was attacked by Bishop Samuel Wilberforce. Upon being asked if he claimed descent from an ape, Huxley created an uproar by declaring that he would sooner be descended from an ape than from a man who misused his position to attack a theory he did not understand.

Modern research has shown, however, that the conventional account of the Huxley-Wilberforce confrontation is a myth.[7] Most of the audience did not feel that Huxley had won the day, and a speech by Hooker probably did more to support the Darwinian cause. Huxley's real role in the Darwinian revolution was to work behind the scenes to ensure that Darwin's supporters gradually came to dominate the scientific community. This was a time in which science as a profession was only just beginning to stand on its feet, and Huxley was determined to throw off the old shackles of subordination to theology. To Huxley, evolution was important because it discredited the kind of science done by the older generation who had accepted the idea of a divine plan of creation. Darwinism was a convenient symbol of the new independence claimed by scientists, and it gained support as the scientific community was taken over by the rising generation of professionals.

The limited role played by the actual arguments and evidence supplied by Darwin is evident from Huxley's lack of response to the detailed program outlined in the *Origin of Species*. He never accepted natural selection as an adequate mechanism of evolution, preferring to believe that species were driven to evolve in certain directions by inherent biological forces. Selection was important not because it was the right answer, but because it showed that scientists could devise new hypotheses with which to open up this fundamental area of investigation. For several years Huxley failed to use even the general idea of evolution in his work on the fossil record. Only in the late 1860s did he begin to stress the discovery of "missing links" to fill the gaps in the record of evolution. This transformation in his interests was sparked not by Darwin, but by the German evolutionist Ernst Haeckel, who developed a comprehensive view of the history of life on earth very much at variance with Darwin's more cautious approach.

There was, of course, a great deal of opposition to the theory of natural selection from conservative scientists. They shared the prevailing belief among religious thinkers that a haphazard process of selection from random variations could never produce the complex adaptations of living things. Selection might weed out a few misfits, but it could never be a truly creative process. The astronomer Sir John Frederick William Herschel was said to have called natural selection the "law of higgledy-piggledy." The anatomist Richard

Owen summed up the feeling that the development of life must unfold according to purposeful direction. In 1868 he contrasted his own theory of "derivation" with Darwin's mechanism: "Derivation holds that every species changes in time, by virtue of inherent tendencies thereto. 'Natural selection' holds that no such change can take place without the influence of altered external circumstances. 'Derivation' sees among the effects of the innate tendency to change irrespective of altered circumstances, a manifestation of creative power in the variety and beauty of the results."[8] A whole host of arguments against the creative power of natural selection were included in works such as St. George Jackson Mivart's *Genesis of Species* of 1871, some of which are still being repeated, virtually unchanged, by the modern creationists.

Unlike creationists, however, Owen and Mivart accepted the basic idea of evolution. It was the mechanism of natural selection they objected to. Both subsequently made important contributions to the debate over the origin and development of new forms of life. In some respects, it is difficult to distinguish their position from that of a Darwinist such as Huxley, who was also inclined to minimize the role of selection. The gulf between the Darwinians and their opponents was symbolic rather than real. The opponents saw Darwin as a figurehead who highlighted the materialism that Huxley and his supporters wished to impose on the discussion. The actual differences between their interpretations of evolution were minimal; it was just that where one party saw the unfolding of a divine plan, the other saw the operations of natural law. By the 1870s the use of quasi-religious language by scientists was on the way out, and to this extent Huxley and the Darwinians had won the day. But the interest in non-Darwinian mechanisms of evolution did not die away.

In America, the *Origin of Species* was attacked by the eminent naturalist Louis Agassiz, although his commitment to an extreme form of creationism weakened his case. Darwin's theory was defended by the botanist Asa Gray, who used arguments based on the distribution of North American plants. Later on, the paleontologist Othniel C. Marsh used fossils discovered in the American West to argue for evolutionism. He stressed the progressive character of evolution by pointing out that there was a steady increase in brain size within most mammalian groups over the course of time. By the 1870s, most American scientists had accepted evolutionism, but as in Britain there were those who had reservations about the implications of the selection theory. Gray was a deeply religious man, and he at first attempted to argue that natural selection was compatible with the belief that evolution was an expression of the divine will (see chapter 3). He later conceded that variation might be led in beneficial directions by a mysterious force. Some of Agassiz's disciples accepted evolutionism, but turned to a Lamarckian interpretation of how the process would work.

In Germany, Darwinism was welcomed by radical scientists, such as Ernst Haeckel, who saw evolutionism as part of a program for reforming society as well as science. While calling himself a Darwinist, Haeckel stressed the progressive nature of evolutionary change and accepted a substantial component of Lamarckism. His evolutionism included a strong element derived from the mystical nature-philosophy which had been popular in early nineteenth-century Germany. Nature was seen as a unified whole with a built-in purpose that would now be expressed through progressive evolution. Haeckel's efforts to reconstruct the history of life on earth became one of the main lines of research within evolutionary biology. For a while this evolutionism was very popular, and it has been said that Germany became the true home of Darwinism. Some of Haeckel's colleagues, including Carl Gegenbaur, accepted the general idea of evolution but were wary of his enthusiastic progressionism. Others biologists were actively hostile. Rudolph Virchow saw Haeckel's evolutionary philosophy as a threat to the social order, and he challenged the scientific basis of the theory as a means of trying to exclude it from the educational system. By the end of the century, Germany too had its anti-Darwinian evolutionists who looked for something more than the natural selection of random variation as the cause of change.

French scientists paid little attention to Darwin's book and were certainly not converted to evolutionism by it. The conversion of the French scientific community did not occur until the 1880s, and then the stimulus was a revival of interest in Lamarck's theory. French culture, based on the rationalism of philosophers such as Descartes, found it difficult to accommodate the haphazard nature of evolutionary change as depicted by Darwin's theory. There was thus no "Darwinian revolution" in France, a fact that should force us to reconsider both the supposed internationalism of science and the assumption that new theories are accepted solely on the strength of the evidence presented in their support. Clearly the scientific arguments for evolution were not strong enough to make a universal impression on biologists. The cultures within which the scientists functioned must have played a role in disposing them to accept or reject the arguments and evidence presented in favor of Darwin's theory.

## The History of Life

By the 1870s everyone in the scientific community had accepted evolution, and the initial phase of the "Darwinian revolution" was over. But Darwin's selection theory remained highly controversial, and the first generation of "Darwinians" made very little use of it. For this reason it is necessary to go beyond the conventional histories of evolution theory to understand what "Darwinism" meant in the late nineteenth century. Because the goal is to highlight the story of how *modern* Darwinism emerged, the orthodox histories concentrate on the story of natural selection, and according to them the key events in the decades following the revolution were those that led toward the development of the modern view of heredity, especially the emergence of Mendelian genetics and its eventual synthesis with the selection theory. This obtrusive use of hindsight can only be misleading, however, for those who wish to understand the contemporary impact of evolutionism. If the debate over heredity was only one, and by no means the main, concern of evolutionary biologists, we shall be left with a distorted view of Darwinism's broader influence if we fail to take account of those areas of biology that were seen as the theory's major areas of application.

The general concept of evolution was exploited most vigorously by biologists who wanted to reconstruct the history of life on earth. Darwin discouraged this grandiose project, and the kind of research that he himself performed was peripheral to the interests of most biologists. Darwin devoted the last years of his life to investigating the evolutionary origins of plant adaptations. Having popularized the view that the history of life could be represented as a branching tree, in which living species represent the tips of the branches, he did not participate in the great effort made by his contemporaries to reconstruct how the branches of the tree fit together. Darwin was always more interested in detailed questions relating to the interactions between species and their environment. The first generation of evolutionary biologists—Huxley included—turned their backs on this kind of research. They wanted to know how the different forms of life were related to one another, and to investigate this they looked at the internal structures of living things, not at their relationships with the outside world.

The classic expression of this form of evolutionary biology was morphology, the science of organic form. Throughout the nineteenth century, biologists had investigated the internal structure of different organisms with a view to expressing the "relationships" between different forms of life. Originally, this was a search to uncover the pattern of divine creation, but Darwin taught that "relationship" had to be understood in natural rather than supernatural terms. Forms were related because they shared a common ancestry, and it was possible, at least in principle, to use comparative anatomy and embryology as a means of assessing the degrees of relationship between all the members of the animal kingdom. Darwin himself had done detailed morphological work on the barnacles, but he was suspicious of attempts to reconstruct the whole evolutionary tree, believing that too much of the evidence was obscured by the imperfection of the fossil record. Others did not share this caution, however, and evolutionary morphology became an active area of research in the 1870s and 1880s.

One of the pioneers of this approach was the German Darwinist Ernst Haeckel. Haeckel saw the reconstruction of life's history as the foundation for a whole evolutionary philosophy that would replace traditional religion. His popular works were translated into English and included *The History of Creation* and *The Evolution of Man*. Huxley was inspired by Haeckel to begin his own researches on the history of certain groups, while his protégés Francis Balfour and Edwin Ray Lankester built their careers on this kind of research. It is important to note that none of these morphologists needed to make use of the selection theory; indeed, they seldom paid much attention to the superficial details of how organisms were adapted to their environment. They wanted to uncover fundamental similarities of structure between different groups. Haeckel, although he called himself a Darwinian, openly favored the Lamarckian mechanism of the inheritance of acquired characteristics.

Much of the morphological work was extremely technical, yet this attempt to reconstruct the history of life produced certain images and metaphors that had the power to influence the wider perception of evolutionism. Haeckel was determined to picture evolution as an essentially progressive process: for him the tree of life had a central trunk that ran straight through to the human race at the top, as the pinnacle of creation. Everything else was a side branch, of lesser importance. It was Haeckel who popularized one of the most powerful images of evolution, the comparison between the history of life on earth and the development of the human embryo. His "recapitulation theory" proclaimed that the whole history of the species is repeated in the development of the modern individual. This was a powerful tool for those embryologists who wanted to study evolutionary origins, but it projected an image of evolution as an inherently progressive, developmental force that would inevitably advance toward ever-higher states of development.

To avoid the idea that the whole process was predesigned by God, the evolutionists visualized life as a force that was constantly challenged and stimulated by the environment. Progress occurred because individual animals strove to dominate their surroundings and then passed their self-improvements on to their descendants. In 1880, E. Ray Lankester stressed the implications of this view by pointing out that any species that moved into a less stimulating environment degenerated to a more primitive level of organization.[9] The embryologist E. W. MacBride, a prominent exponent of both Lamarckism and the recapitulation theory, argued that the invertebrates were merely branches of the main vertebrate stem of evolution that had turned aside and degenerated through the adoption of a less stimulating way of life. He wrote that "the Invertebrates collectively represent those branches of the Vertebrate stock which, at various times, have deserted their high vocation and fallen into lowlier habits of life."[10] The idea of progress through struggle could thus be uncoupled from the selection theory with the insistence that the struggle was primarily against the environment.

The fossil record was also of great potential importance in the reconstruction of the history of life. Darwin had warned that the record was so imperfect that little could be gained from it, but the following decades saw a vast explosion in the amount of fossil evidence available. Huxley played a major role in the interpretation of some discoveries: he stressed the importance of the fossil *Archaeopteryx* as a missing link between reptiles and birds, and hailed a series of American fossils uncovered by O. C. Marsh as "demonstrative evidence" of the evolution of the horse. By the end of the century, it was possible to chart the origin and history of the various families of mammals with some accuracy. At this level, the development of life could be seen as a series of distinct episodes:

each family went through a cycle of growth, maturity, and decline, often ending in extinction as some newly developed form came in to take its place.

Although this model of history stressed the competition between different groups, it had little use for the theory of natural selection as Darwin had conceived it. Indeed, most paleontologists preferred Larmackism to explain the development of adaptive characters. Among them was the paleontologist Edward Drinker Cope, who led a prominent group known as the American neo-Lamarckians. Cope also stressed the development of what he believed to be non-adaptive characteristics in many families of animals. He thought that each group went through a cycle of development paralleling the life history of an individual, from birth, through maturity and senility, to death. Advocates of this theory of "othogenesis" believed that inherent forces drove the evolution of each group in a particular direction, even toward the eventual production of harmful characters. While accepting that the overall history of life was progressive, paleontologists saw the process as a series of cycles corresponding to the rise and fall of the successive groups that have dominated the earth. Here was a powerful source of metaphors for parallel accounts of the history of human civilization.

One of Darwin's own interests that did eventually take off as a research program was biogeography, the study of the geographical distribution of species. It was this topic that had first led Darwin to accept evolution, and he had entered into a long debate with Hooker on the means by which newly evolved species might spread out to occupy more territory. In 1876 Wallace published a massive study, *The Geographical Distribution of Animals*, that stimulated much debate over the following decades. Increasing numbers of biologists took up the attempt to reconstruct the history of life's migrations around the globe. It was possible (especially when fossil evidence became available from different continents) to reconstruct the history of each group, its point of origin, and its subsequent migrations to new territories.

Wallace pioneered the view that the harsh Eurasian climate had provided the stimulus that had led to the successive evolution of higher types of life. Even the human race was supposed to have evolved there before migrating around the world. The details of animal distribution could be explained by invoking geographical barriers that had sprung up at various points in geological time to block the migration of newly evolved forms spreading out from Europe and Asia. The marsupials of Australia, for instance, had been protected by the emergence of a sea barrier, which saved them from extermination by the more highly evolved placental mammals. This model of successive waves of expansion from Eurasia reached its apotheosis in William Diller Matthew's paper "Climate and Evolution" of 1914.[11]

Biogeography has received little attention from historians of science, yet it is crucial for an understanding of the transfer of metaphors between biology and ideology. Without invoking the theory of natural selection, the Wallace-Matthew approach provided a perfect reflection of imperialism in biology. Northern Eurasia, classically the home of the white races, was seen as the center of evolutionary progress. A stimulating environment was invoked as the cause of progress, so that any species (or race) pushed out of the center was doomed to stagnation. The history of life was pictured not as a continuous progression, but as a series of distinct episodes in which a succession of higher forms radiated outward, exterminating most of their predecessors.

## The Debate over Selection

None of these competing models of history depended exclusively on Darwin's selection theory. Natural selection remained controversial well into the twentieth century, and its eventual rise to dominance marks a second phase of the "Darwinian revolution" that must be taken into account if we are to understand the modern implications of Darwinism. The critical question was the origin and inheritance of the adaptive characters that were supposed to be picked out by natural selection: Darwin himself proposed a model of variation and inheritance that modern geneticists dismiss as simply wrong. Yet his theory was characteristic of mid-nineteenth-century thought on these issues, and its replacement by modern genetics represents a fundamental breakthrough in biological thought. Some of the later implications of selection attributed to Darwinism certainly depend on the new model of heredity, and it is important to be aware of the differences between the old and new versions of the theory.

Darwin's own hypothesis of heredity was called "pangenesis." It implied that the offspring is a simple amalgam of its parents' characters. It also allowed for the inheritance of new characteristics ac-

quired by the parents' efforts—and in this Darwin accepted a subsidiary element of Lamarckism. The pregenetical character of much
mid-nineteenth-century thought on heredity created a climate of
opinion within which the Lamarckian alternative could flourish.
More problematic for the selection theory was the concept of
"blending heredity." As the Scottish engineer Fleeming Jenkin
pointed out in 1867, natural selection would not work if an important new characteristic appeared in a single individual.[12] That individual would have to breed with unchanged members of the
population, thus diluting the new characteristic by half. Over the
next few generations the new characteristic would gradually
disappear, swamped by interbreeding with the rest of the population. According to the conventional historiography of Darwinism,
Jenkin's argument demonstrated that natural selection would only
work with a particulate model of heredity, that is, one in which the
characteristics are transmitted as indivisible units from one generation to the next, as in Mendelian genetics.

The conventional interpretation is wrong on this score, however,
because it was quite possible to develop a plausible model of selection based on blending heredity. The Darwinians of the late nineteenth century did not assume that new characteristics appeared in
single individuals; they saw selection acting on a wide range of
variation existing naturally in all populations. Nevertheless it was
clear that the selection theory would require a satisfactory mechanism of heredity if it were to be taken seriously, and two schools of
thought independently tried to tackle this problem. Darwin's
cousin Francis Galton openly criticized pangenesis and began to insist that heredity is a much more rigid process by which characteristics are transmitted from one generation to the next. He was
particularly interested in heredity acting within large populations,
and in developing statistical techniques that could be used to test
the effect of mechanisms such as selection on a varying population.
Building on Galton's foundation, his disciple Karl Pearson developed a thoroughly selectionist model of evolution during the 1890s.

For Galton, the individual's character was rigidly determined by
inheritance. Improvements as the result of a better environment or
(in human beings) of education were not possible. The social implications of this hereditarian view of human nature are an important
component of "social Darwinism"—yet it must be stressed that until the last decades of the nineteenth century hardly anyone saw
heredity as having such as deterministic role in human character.

The true nature of this new concept of heredity was brought out even more clearly in the work of the German biologist August Weismann in the 1890s. Unlike Galton, Weismann was interested in the cellular basis of heredity, and he correctly realized that the chromosomes of the cell nucleus are crucial for the transmission of inherited characteristics. Weismann postulated that the chromosomes were composed of a substance that differed from the rest of the body, and that this substance, the germ plasm, somehow stored the information necessary to build the body of a new organism. He believed that the germ plasm was transmitted from one generation to the next through the cell nucleus, independently of the rest of the parent's body. Changes to the parent's body due to nutrition, exercise, and the like were not reflected in the germ plasm by which reproduction takes place.

Weismann thus declared that Larmarckism was nonsense: acquired characteristics could not be inherited, and natural selection was the only conceivable mechanism of evolution. This theory formed basis of the "neo-Darwinism" of the 1890s, a form of Darwinism purged of the non-Darwinian elements that Darwin himself had been unable to throw off. Weismann's theory of the germ plasm's structure was still very speculative, however, and it bore little resemblance to the modern conception of the gene. His views merely provoked those who still thought that Lamarckism was an essential component of evolutionism into an outright rejection of Darwinism in its new form. Herbert Spencer, a supporter of the old form of Darwinism, now joined with critics such as Samuel Butler to attack Weismann's position. Neo-Darwinism so polarized opinions that neo-Lamarckism emerged as a distinct school of thought opposed to Darwinism.

Another line of opposition to Darwinism was emerging among naturalists who thought that the appearance of new species was a discontinuous process, that new forms appeared suddenly as the result of an instantaneous transformation of the hereditary constitution of an organism. This position was advocated by William Bateson in his book *Materials for the Study of Variation*, published in 1894. Originally an evolutionary morphologist, Bateson had come to reject the whole apparatus of traditional Darwinism, declaring that natural selection had nothing to do with the production of new characteristics. The belief that traits appeared as spontaneous new "units" stimulated interest in the possibility that the existing character of a species might be a mosaic of individual units, not a continuous spectrum as the Darwinians supposed. Partly in response to this suggestion, several biologists began conducting breeding experiments in the hope of identifying such unit characters. In 1900, Hugo De Vries and Carl Correns independently produced evidence for this position, and then they found that laws governing the inheritance of unit characters had already been proposed more than thirty years earlier by Gregor Mendel. Mendel's classic experiments of the 1860s had gone largely ignored, but now they were hailed as the foundation of a new theory of heredity.

Bateson soon took up the new theory, and it was he who coined the term "genetics" to denote the study of the inheritance of unit characters. But both Bateson and De Vries still insisted that new species were formed instantaneously through the production of new genetic units, by the process that De Vries called "mutation." There was no role seen for the selection of small character differences, only for the sudden appearance of major new features marking off distinct, specific types. Genetics provided a new conception of the nature of Weismann's germ plasm and was similarly hostile to Lamarckism. Bateson was especially determined to root out the last vestiges of Lamarckism in science. He was responsible for discrediting the Lamarckian experiments of the Austrian biologist Paul Kammerer in what has become known as "the case of the midwife toad," after the subject of one of Kammerer's most controversial tests.[13]

Genetics gradually eliminated Lamarckism from biology, especially in the English-speaking world. The eclipse of Lamarckism was less rapid elsewhere, and in Stalin's Russia it enjoyed a revival under the regime of the plant breeder T. D. Lysenko. Lysenkoism is usually dismissed as a scientific disaster caused by explicit political interference, and although the true story is rather more complex, it is fair to say that Lamarckism has never succeeded in reversing the trend against it within science—despite a series of attempts to restore its credibility.

The hostility of the geneticists to Lamarckism did not, at first, translate into support for Darwinism. The Darwinians were committed to the concept of selection from a range of continuous variation and were suspicious of the new emphasis on discontinuous character differences. Karl Pearson had already attacked Bateson's position, and was now forced to reject Mendelian genetics as an extension of the same anti-Darwinian view of evolution. In principle, the concept of unit characters might have helped Darwinism, since new units could not be diluted by interbreeding. But recognition of this point was impossible so long as the two sides had such widely divergent views on the nature of variation.

In the 1920s a reconciliation between the two positions was gradually worked out. Pearson's student R. A. Fisher realized that many Mendelian units might be so small that, when interbred within a large population, they would give the appearance of a continuous range of variation, just as the Darwinists assumed. Selection would consist of the picking out of those genes that conferred adaptive advantage, and new genes would be introduced into the population by mutation. By modifying Pearson's statistical techniques, Fisher and others showed how the genetic structure of a population would gradually change when subject to selection. There was no need for the geneticists to think of the production of new species as a sudden process, because large-scale mutations would not participate in the natural process of evolution. Fisher's *Genetical Theory of Natural Selection*, published in 1930, helped to consolidate the new form of Darwinism.

The new Darwinian form of population genetics was so mathematically complex that at first few field naturalists could understand it. But some biologists studying the geographical variation of populations were independently beginning to realize that the Darwinian approach had much to offer. In 1937 Theodosius Dobzhansky's *Genetics and the Origin of Species* demonstrated how the mathematical theories of the population geneticists could be translated into predictions about natural processes in the wild. Naturalists such as Ernst Mayr and Julian Huxley now began to stress the importance of the new Darwinism, with Huxley's *Evolution: The Modern Synthesis* of 1944 helping to create the impression that here was a new theory uniting the best of the old traditions. In the same year G. G. Simpson's *Tempo and Mode in Evolution* showed how paleontology—once a hotbed of anti-Darwinian thinking—could be incorporated into the new approach. By the 1950s the modern-synthetic form of Darwinism had begun to gain the dominant position in biology that it has enjoyed ever since. Many efforts to challenge or modify the basic Darwinian position of the theory have been made, but none has succeeded in toppling it from its pedestal. Modern evolutionary biology is still predominantly Darwinian—but it is a very different kind of Darwinism to that upheld by Darwin's supporters in the 1860s.

# 3

## Darwinism, Religion, and Morality

In the modern world we have been led to believe that Darwinism represents the spearhead of aggressive materialism in its assault on traditional religious values. Creationists blame scientific evolutionism for persuading us that the world has no transcendental purpose; that humans are brutes living in a world where survival and material pleasure are all that count. Many intellectuals who do not share the creationists' commitment to the Biblical story of human origin nevertheless accept that Darwin's theory played a major role in undermining moral values. If evolution is no more than the natural selection of random variations, living organisms—including human beings—are reduced to puppets manipulated by the strings of their genetic inheritance. Inspired by this negative image, historians such as Jacques Barzun and Gertrude Himmelfarb have interpreted the Darwinian revolution as a great turning point ushering in the Western world's cultural decline.[1]

It is easy to see how the controversy sparked off by the *Origin of Species* can be presented as the starting point of a debate that it still going on today. Many religious thinkers of the 1860s greeted Darwin's book with open hostility. According to one story, Darwin was pointed out as "the most dangerous man in England" by a clergyman. The confrontation between T. H. Huxley and Bishop Samuel Wilberforce at the 1860 meeting of the British Association for the Advancement of Science seems to symbolize the clash between a rising scientific naturalism and the traditional values that had sustained Western civilization for centuries. There a man of the church stood up to defend the view that the world was created by a wise and benevolent God, against those who would reduce the story of creation to a chapter of accidents.

The most heated exchange of that famous debate occurred when Wilberforce asked Huxley if he claimed to be descended from an ape on his mother's or father's side. It was the inclusion of the human race in the system of evolution that brought the worries of conservatives to a head. In 1863, Huxley's book *Man's Place in Nature* made it clear that in an evolutionary worldview, humankind was part of nature, not raised above it by the possession of unique spiritual powers. The following year, Wilberforce invited the British politician (and future prime minister) Benjamin Disraeli to Oxford, where he responded to the question, "Is man an ape or an angel?" with the immortal line: "My Lord, I am on the side of the angels."[2] A host of cartoons in the popular press lampooned the evolutionists' claim that the human race was descended from the apes. The most common image of the ancestor was the gorilla, whose allegedly savage character had just been highlighted by the explorer Paul Du Chaillu's accounts of his discoveries in Africa.

It is all too easy to fall into the trap of assuming that the modern creationists' hostility to Darwinism can be linked to the initial response to the *Origin of Species*, as part of a continuous antimaterialist tradition. The inevitability of the "warfare" between science and religion has been stressed by those historians who favor the sweeping aside of outdated superstition. The opponents of materialism also find it convenient to portray the relationship between science and religion as based on conflict. Both sides in the debate have an interest in polarizing our view of the situation by convincing us that scientific theories such as Darwinism necessarily destroy traditional values—they merely disagree over whether to applaud or condemn the destruction. In this view, it was inevitable that religious thinkers would attack Darwin's *Origin of Species* and would continue to resist evolutionism despite its growing popularity among scientists.

Recent historical research has revealed, however, that the war between science and religion is something of a myth.[3] It is an artifact of hindsight, a distorted image produced by applying modern values to a past situation. However strong the initial condemnation was, we now know that religious hostility to evolutionism declined quite rapidly in the late nineteenth century. It was certainly an age of growing secularization, and conservative religious thinkers were having to defend their beliefs on a number of fronts. In fact, the debate over Darwinism coincided with an equally bitter controversy over the status of the Bible as a historical document. But many liberal Christians were able to adjust their beliefs to the new situation so that the idea of evolution could be incorporated into a system that still retained many traditional values.

By the 1870s most educated people had accepted evolutionism—but they had certainly not become atheists, nor even agnostics (a term coined by T. H. Huxley to denote an active refusal to take one side or the other on the question of the existence of God). The conflict was smoothed over because many scientists still wished to believe that the universe was a system with a moral purpose, and because most theologians were prepared to accept that evolution was God's way of allowing His creation to unfold. Teleology (explanation in terms of the purpose to be achieved by a process) was retained in a modernized form. In Paley's argument from design, the purpose of each species was built-in by its Creator, who pre-adapted it to its mode of life. In the new teleology, the purpose was a long-range goal which the Creator intended evolution to achieve. However surprising it might seem to modern creationists, the late nineteenth century saw the erection of an evolutionary compromise that allowed both sides to feel comfortable with the new science. Even some of the early fundamentalists felt able to accept evolutionism.[4] The hostility of the modern creationists dates back only to the 1920s, when many rural Americans sought a scapegoat for what they saw as the decline of traditional values, and began to identify Darwinism as the cause.

To understand this historical seesaw of opinion, it must be remembered that the popular "Darwinism" of the 1860s and 1870s was not the neo-Darwinism of modern biologists. From the start, the opposition of religious and moral thinkers was directed against Darwin's theory of natural selection. They realized—and Darwin himself was increasingly inclined to agree with them—that it was very difficult to imagine that a wise and benevolent creator would use so harsh, wasteful, and selfish a mechanism to achieve His ends. Evolutionism could be acceptable to these critics only if the process could be seen to work in a way that retained a role for divine purpose. Animals (and especially humans) need not be reduced to biological puppets; they could play an active role in directing evolution in a morally significant direction. Progress, meaning *moral* progress, would be the central feature of evolution. But a rigid dependence on Darwinian natural selection would render this view impossible. A compromise was constructed by ignoring or evading the harshest implications of the selection theory to retain a sense of progress and purpose in nature.

There is thus a sense in which the opponents of Darwinism have always attacked the same thing—natural selection—while the supporters at first toned down exactly those features of the theory that modern biologists find most persuasive. If modern biologists are correct in their belief that Darwin was ahead of his time, it was the anti-materialists who were most perceptive in identifying the threat posed by the new theory. As biologists and materialists have become ever more attracted to natural selection as a means of eliminating the old teleology, the opponents of Darwinism have mounted a long-standing and successful campaign to pinpoint the selection theory as a threat to traditional religious beliefs and moral values. Thus it is the modern biologists who have been forced into a situation in which they must defend a theory that has long been labeled as a dangerous manifestation of materialism. Their predecessors were able to adopt a much more flexible policy when defending the original Darwinism, and this difference must be taken into account when evaluating the impact of the theory on Darwin's contemporaries and its effect on late nineteenth-century thought.

Our analysis is complicated by the fact that alternatives to Darwinian natural selection—especially the Lamarckian theory of the inheritance of acquired characters—were eventually brought into the debate on the assumption that they had less unpleasant moral implications. Opponents of Darwinism have often claimed that *evolutionism* is acceptable, provided it operates through a more purposeful process than natural selection. Darwinism is evil because it portrays the world as a scene of purposeless struggle, but Lamarckism is good because it allows living things to progress through their own efforts. Unfortunately for those who would pin these clearly defined moral labels on the theories involved, history suggests that the moral implications of theories depend on the circumstances against which they are evaluated. The early form of "Darwinism" contained strong Lamarckian elements through its absorption into Herbert Spencer's philosophy of universal progress. For Spencer's supporters, this Lamarckian element—which is non-Darwinian by modern standards—allowed evolutionism to present itself as an expression of a modernized version of traditional Protestant values. Later critics have forgotten the role of Lamarckism in Spencer's philosophy in their desire to depict him as an exponent of a ruthless "social Darwinism." Once again, modern preconceptions about the implications of scientific theories distort the interpretation of their influence in the past.

## Darwinism and Design

In the previous chapter I showed how early nineteenth-century ideas about the origin of species were dominated by the view that each form of life had been individually designed by its Creator. William Paley's *Natural Theology* (1802) argued that the adaptation of each species' physical structure to its way of life confirmed the Creator's wisdom and benevolence, thus reinforcing the "argument from design" that had been popular for centuries. But this relatively simple image of the relationship between God and nature broke down long before Darwin published his theory. In an effort to keep the idea of a divine architect up-to-date in terms of the discoveries being made in natural history and the fossil record, biologists such as Richard Owen had begun to argue that the Creator's plan could best be seen not in the details of how each individual species was formed, but in the overall pattern of development that linked all the species into a coherent and meaningful whole. The idea of progress was invoked to give a direction to the history of life: in the course of time the Creator had introduced successively higher forms of life, culminating with the human race. But precisely because this idea could be exploited by radical evolutionists, conservatives such as Owen stressed that the Creator's plan unfolded in a series of distict episodes, not as a continuous trend.

Darwin himself had confronted these issues as soon as he had begun to explore the possibility of constructing an evolutionary theory. Natural selection made use of variation that was random in the sense that it produced a wide range of different characteristics. Hence there could be no preordained direction of evolutionary development, no goal toward which life must inevitably advance. Species changed solely because certain characteristics were selected by the external environment, which was itself constantly changing in an unpredictable manner. Adaptation became a natural process, not a fixed state designed by God, but by its very nature an evolutionary process driven by the demands of adaptation would be unpredictable and, to all intents and purposes, haphazard. Each division and subdivision of the tree of life would move off along its own path, quite unrelated to what was going on in all the other branches. To cap it all, the process of selection meant the elimination of all those individual variations that resulted in nonadaptive characters: suffering and death were inevitable if random variation were to be directed along useful channels.

Historians have long disagreed over Darwin's religious beliefs, but there is now a growing consensus that his ideas changed significantly in the course of his career. Initially he saw natural selection as a process that might still be reconciled with the belief that the universe was designed by God. His was a materialistic theory, but not necessarily an atheistic one. Everything worked in accordance with natural law, but the laws themselves had been instituted for a purpose by a wise God who chose to achieve His ends indirectly rather than by the exercise of supernatural power. Natural selection explained the existence of suffering in the world: it was necessary because only by this means could species adapt to an ever-changing environment. Darwin also hoped that, in the long run, natural selection would tend to improve species by making them more adaptable and more intelligent. There might be no "main line" of evolution aimed at a single goal, but there was a general pressure tending to improve the level of organization in most branches of the tree of life. The conclusion of the *Origin of Species* stressed the view that there might be a long-range purpose built into evolution, and thus encouraged the book's readers to see the theory as an extension of traditional religious values.

In the decade following the publication of the *Origin of Species*, however, Darwin began to have doubts about the viability of this compromise. He became more aware of the relentless pressure of natural selection, and of the suffering that resulted from its essentially selfish activity. It was difficult to see a process that could lead some species to gain their livelihood by acting as parasites on others as the means of evolution preferred by a benevolent God. Natural theologians had explained predators as God's way of giving sick and aged animals a quick death, but the parasite lives by causing suffering—its mode of life is the ultimate expression of nature's inherent selfishness. The parasite also degenerates as it adapts to a less-active lifestyle, thus confirming that there is no progressive trend in evolution. Although still tempted by the idea of progress as a means of giving a direction to evolution, Darwin's biological studies made him increasingly aware of how difficult it was to compare developments in one branch of evolution with those in another, thus rendering any notion of a natural hierarchy untenable. In his later years, Darwin tended to an agnosticism that vacillated between a vague hope that evolution might eventually be reconciled with belief in God and a pessimistic fear that no purpose whatsoever could be discerned in the history of life.

For religious thinkers imbued with the spirit of Paley's natural theology, the theory of natural selection posed major difficulties. It has been said that Darwin stood Paley on his head, and there were certainly obstacles standing in the way of any attempt to see selection as a mechanism instituted by a benevolent God. Evolution by some other means might be seen as the natural unfolding of a divine plan—but surely a God who cared about His creation could have thought of a mechanism less wasteful and less horrific than the constant elimination of the unfit. Everyone conceded that there were occasional errors in the reproductive process that gave rise to monstrosities, and no one objected to the idea that these would normally be eliminated. A kind of selection was, in fact, necessary to prevent the species from being contaminated by such harmful characteristics. But to suppose that reproduction was *intended* to generate a wide range of variant characteristics, most of which would turn out to be useless or harmful, was to build a rejection of teleology into the most basic of biological processes. Selection was acceptable as a negative process for eliminating nature's rare mistakes, but few could see it as a positive factor that was essential for turning the variability of life to any account. The actual production of fitter characteristics was the crucial issue: were they merely the lucky few out of a range of otherwise useless forms, or was there something in nature that directed species to produce the traits they needed to evolve?

This problem was addressed by one of Darwin's chief supporters in America, the botanist Asa Gray, who was a deeply religious man. In a series of articles later collected in his *Darwiniana* of 1876, Gray at first tried to argue that evolution itself had no more implications for theism (the belief that the universe is governed by God) than the laws of physics. Understanding how the laws of nature worked told us nothing about how those laws were instituted, Gray believed. Any process of adaptation fulfilled a purpose by allowing species to maintain themselves in an ever-changing world. Darwin would surely not want to claim that a complex structure such as the eye was the result of trial and error, Gray argued, in which case it had to be seen as the intended result of a purposeful evolutionary process.

In the end this line of reasoning led Gray to question the validity of the selection mechanism. He claimed that Darwin would save himself much trouble if he abandoned the idea of random variation, because then there would be no "scum of creation" (nonadaptive variants) to be eliminated in every generation. Nature does not produce "crude, vague, imperfect and useless forms," he wrote, and Darwin would be advised to assume instead "that variation has been led along certain beneficial lines."[5] Darwin realized that this assumption undermined the whole logic of his theory: if variation drove evolution in a preordained direction, there would be no need for selection. In any case, his own observations on domesticated species confirmed that most individual variations *were* useless. In the conclusion of his *Variation of Animals and Plants under Domestication,* Darwin answered Gray directly, using the following analogy to make his position clear. If a man built a house out of stones that had fallen naturally from a cliff, picking out those that were of a useful shape and rejecting the rest, we would hardly conclude that the physical processes that eroded the cliff were designed to help the builder. The design came in the selecting process, not in the production of the raw material. Struggle and suffering would be pointless if they did not serve the purpose of eliminating the unfit variants.[6]

Suffering was necessary for adaptive evolution—but did this mean that God *intended* the suffering to occur? Darwin confessed himself unable to resolve this dilemma, but he was adamant that variation was not a purposeful trend that could be said to represent a divinely implanted creative force. To most of his contemporaries, however, it seemed that selection was a purely negative process: it eliminated the unfit but did not explain the production of the fit. Given that the cause of variation was unknown (in the period before the discovery of genetic mutation), would it not be better, they thought, to follow Gray's assumption that there were purposeful forces built into nature that could guide evolution in certain directions and that could thus be seen as expressions of a divine purpose? Such misgivings about the logic of natural selection were perhaps to be expected among religious thinkers, but many scientists also found it difficult to cope with the idea of a universe that was governed largely by chance. The central plank of the movement sometimes called "scientific naturalism" was that nature was governed by law, not by arbitrary miracle. But precisely because scientists wanted a law-bound universe, they were unwilling to go along with Darwin's theory. Natural selection was a combination of law-bound processes, but it was a combination so complex that the outcome in any particular case was unpredictable. To be acceptable, these scientists believed, evolution would have to be based on a law of development that drove it in a predetermined direction.

As noted in chapter 2, the prominent astronomer Sir John Frederick William Herschel dismissed natural selection, calling it the "law of higgledy-piggledy." Herschel insisted that "an intelligence, guided by a purpose, must be continually in action to bias the direction of the steps of change—to regulate their amount—to limit their divergence—and to continue them in a definite course."[7] Richard Owen, a vitriolic critic of Darwin and Huxley, abandoned his outright opposition to evolution and proposed a theory of the "derivation" of species, in which there was "an innate tendency to change irrespective of altered circumstances" that could be seen as "a manifestation of creative power in the variety and beauty of the results."[8] Another comparative anatomist, St. George Mivart, aroused the enmity of the Darwinians with his book *Genesis of Species* (1871), in which he argued that evolution was far too regular and orderly a process for it to be the result of random variation and haphazard changes in the environment. The world was the product not of trial and error, Mivart claimed, but of a rational divine mind that had built preordained trends into nature.

Herschel, Owen, and Mivart were arguing for what has sometimes been called "theistic evolutionism"—the belief that evolution is somehow controlled by a supernatural power that guides it in a purposeful direction. By this means, they were able to reconcile the basic idea of evolution with their religious values. Yet in an age when materialism was gaining ground, it was difficult to defend the claim that supernatural forces were constantly at work, interfering with natural processes. The American philosopher John Dewey would eventually ridicule this theory as "design on the installment plan."[9] What was needed was a process that seemed compatible with the idea of divine benevolence, yet that could be seen as purely natural in its operation. There was indeed such a process available in the era before the advent of Mendelian genetics: Lamarck's long-despised theory, the inheritance of acquired characteristics. If one could believe that traits acquired in response to animals' purposeful behavior were inherited, then these inherently adaptive modifications would take the place of Darwin's random variations and would guide evolution along beneficial lines. There would be no need for the elimination of the unfit, because all the animals would become fitter as they adjusted their behavior to changed circumstances.

Lamarckism—once dismissed as a radical, materialistic theory—now seemed to offer a haven of refuge for those alarmed by the far more materialistic implications of natural selection. One of the first to recognize this possibility was the novelist Samuel Butler, who began as a supporter of Darwinism, but was alerted to the dangers of selectionism by Mivart's attack. In his book *Evolution Old and New* of 1879, Butler criticized Darwin for failing to acknowledge the importance of earlier evolutionists such as Lamarck, and he assailed natural selection as a materialistic theory that reduced living organisms to the status of automatons. For the selectionist, the organism is totally passive: it lives or dies according to the value of what we would now call its genetic inheritance. Nothing the organism can do will alter its fate, and most will, in fact, die out because they possess nonadaptive characters. As Butler wrote in a later attack: "To state this doctrine is to arouse instinctive loathing; it is my fortunate task to maintain that such a nightmare of waste and death is as baseless as it is repulsive."[10]

In the Lamarckian theory, however, organisms play an active role by adapting themselves to new situations, as when the ancestral giraffes adopted the habit of feeding from trees. Such purposeful behavior changes the animals' physical traits and—when inherited—affects the future evolutionary development of the species. Evolution is purposeful because it is guided by the purposeful behavior of the animals themselves, and there is thus a sense in which this behavior partakes of divine creativity: God has delegated His creative powers to life itself, endowing all living things with the capacity to shape evolution. Many Lamarckians were, in fact, "vitalists"—they believed that living things were driven by a purposeful force that could not be reduced to the laws of physics or chemistry.

Butler developed this theme in a series of books that included violent personal attacks on Darwin. In response he was ostracized by the Darwinians—yet by the last decade of the nineteenth century, his views were being openly supported by a number of British scientists, including Darwin's son, Francis. As Darwinism hardened into neo-Darwinism, an increasing number of scientists and nonscientists found that they were unable to go along with rigid selectionism and turned instead to Lamarckism. In America an active school of neo-Lamarckism had been established in the 1870s, led by the paleontologist Edward Drinker Cope, another deeply religious scientist who openly defended his views. Cope entitled a collection of essays *The Origin of the Fittest* to stress that Lamarckism addressed the question of how useful characters are produced, not merely how they survive. He thought that he had good evidence from the fossil record that evolution was guided in beneficial directions and thus could not be the product of trial and error. Like Butler, he saw Lamarckism as a theory that was both scientifically and morally preferable to Darwinism, a view he developed in his book *Theology of Evolution*, published in 1887. He too saw the active behavior of living things as a manifestation of God's purpose. All animals, he believed, possess some degree of consciousness, and this mental power expresses an underlying element of divine creativity implanted in nature to guide evolution. The Lamarckian process also ensured that evolution would be progressive, because animals would use their intelligence more in each succeeding generation.

In the early twentieth century Lamarck's theories were disproved and eliminated from biology, but among intellectuals Lamarckism retained its reputation as a morally preferable alternative to Darwinism. The playwright George Bernard Shaw supported it in the preface to *Back to Methuselah* (1921), declaring, "If it could be proved that the whole universe had been produced by . . . selection, only fools and rascals could bear to live."[11] More recently, the author Arthur Koestler wrote a series of books, including *The Ghost in the Machine* (1967), attacking Darwinism as an example of the materialism that was destroying our respect for life, and offering the prospect that Lamarckism might yet be found to have a grain of truth in it. As a result of efforts by these literary figures, many still believe that Darwinism is an inherently evil philosophy of nature and that non-Darwinian mechanisms such as Lamarckism must play a role in any wholesome and humane worldview. Unfortunately for those who would offer such a black-or-white choice, history shows that Lamarckians have been just as supportive as Darwinians of objectionable social philosophies such as racism (see chapter 4). Whatever the theological problems arising from the selection theory, the attempt to use moral values as the basis for choice between scientific theories founders on the complexity of the issues involved.

## Man's Place in Nature

The debate over the theological and moral implications of the evolutionary process was all the more heated because it was obvious that, according to the new philosophy, the human race would have to be regarded as a product of this process. Instead of being set apart from the rest of the animal kingdom by virtue of having a spiritual dimension, people would have to regard themselves as being merely highly developed animals. Darwin avoided discussion of human evolution in the *Origin of Species*, but it was clear from earlier debates that this was a crucial issue. Huxley and Owen were already disagreeing over the degree of anatomical similarity between humans and apes, and Huxley's *Man's Place in Nature* of 1863 presented the evidence for a close link, although it did not explore the actual process of human evolution.

Also published in 1863, Charles Lyell's *Antiquity of Man* popularized a new interpretation of archaeology, in which it became apparent that a primitive, Stone Age human culture had been on the earth for a vast period of time. Although geologists had long realized that the earth itself was much older than the Biblical creation story implied, it had remained an article of faith that the human race had appeared only a few thousand years ago. Now irrefutable evidence had emerged that the human race was of much greater antiquity, and it was equally clear that our distant ancestors had lived in a state equivalent to that of the lowest modern "savages." Although some critics tried to argue that a Stone Age technology did not necessarily imply that our ancestors were *morally* inferior to ourselves, most anthropologists and archaeologists assumed that Stone Age culture had been extremely limited. Anthropologists postulated a ladder of cultural development leading from primitive hunter-gatherers through to modern industrial civilization. It was inevitable that the idea of cultural progress would soon be linked to the theory of biological evolution. The Stone Age ancestors (and, by implication, modern-day primitives) were seen as intermediate steps in the ladder of intellectual and moral development leading from the apes to the modern civilized humans. In his book *Prehis-*

*toric Times,* published in 1865, John Lubbock linked the development of human culture to a general philosophy of progress, presenting the triumph of civilization as the inevitable outcome of the progressive development of life depicted by the new evolutionary world view.

Conservative thinkers inevitably found this downgrading of humankind's status unacceptable. But it was not only conservatives who balked at humanity's wholesale inclusion in the natural world. To begin with, many scientists also found it difficult to believe that all aspects of human character could have been derived by natural processes from the mentality of animals. Lyell himself thought that mankind must have been produced by an evolutionary leap or saltation in which the characteristics separating us from the animals were suddenly introduced. Alfred Russel Wallace, the codiscoverer of natural selection, became convinced that many aspects of the human mind could not have been generated by natural selection, and he argued openly for supernatural intervention in the later stages of human evolution. Such doubts would only be overcome if people were convinced that the evolutionary process was capable of forcing life to progress up to the human level, and it would be much easier to do this if the evolutionary process were depicted as an inherently progressive and morally purposeful force.

In the end, an evolutionary account of human origins was accepted because the more radical implications of the selection theory were ignored or evaded by the early Darwinians. By implying that evolution necessarily progressed up to the human level of intellect and moral awareness, the evolutionists were able to argue that it was no denigration of the human species to suppose it to be the product of such a purposeful system. The idea of progress was an essential component of Victorian evolutionism precisely because it played a vital role in making the theory acceptable to those who feared a total destruction of the traditional view that the human race plays a central role in a morally significant universe. In effect, the human characters and values that were traditionally supposed to separate us from nature were transferred into nature itself by claiming that evolution was a process with a built-in moral goal.

Darwin addressed the question of human origins in *The Descent of Man,* published in 1871. It was the first detailed attempt to explain how the human race could have been produced by purely natural causes. Darwin tended to minimize the mental gulf between humans and the highest animals by attributing almost-human behavior to apes, dogs, and other species. Anecdotal evidence was used to argue that these animals had significant levels of intelligence and were even capable of altruism in their relationships with others. But the important question was, Why had the human species developed a much higher level of mental and moral activity than any other species? Most of Darwin's contemporaries assumed that evolution inevitably progressed up a scale of mental development toward the human level. Given time, it was to have been expected that humans—or at least creatures with human levels of mentality and human moral values—would appear on the earth.

For Darwin himself, however, this assumption of a ladder or scale of development was not an acceptable mode of explanation. He was quite capable of following the progressionist views of his time when portraying the white race and European civilization as the goals of human development. But when it came to explaining the origins of the entire human race, he realized that with a branching model of evolution it was necessary to ask why the human branch had developed in a different direction from that of its closest relatives. His answer was to invoke a change of habitat in which our distant ancestors left, or were driven out of, the forests, while the ancestors of the apes remained living in the trees. In the new habitat, it was an advantage to stand upright, and this freed the hands from the duties of locomotion (climbing) and left them free to become organs of manipulation. From this followed the use of tools, and it was this ability to explore and take control of the environment that stimulated the development of intelligence. Darwin believed our ancestors stood upright before they developed human levels of intelligence.[12] This was a theory that would not be confirmed until the twentieth century, and it implied that the appearance of the human race was not predictable from the general laws of evolution. Without the unique combination of circumstances that led to our ancestors leaving the trees (what modern evolutionists call an "adaptive scenario"), the human race would not have evolved.

Darwin assumed that human moral values were the product of mankind's superior level of intelligence acting on the instincts that evolution must imprint on all social animals, to make group living possible. He invoked Lamarckism to explain how individuals would learn the advantages of cooperative behavior and then transmit this knowledge as inherited instincts to their offspring. He also supposed that natural selection would operate between groups, with the less cooperative groups being eliminated by their rivals. Thus cooperative and altruistic behavior would gradually be enhanced in those species that lived in groups. When humans developed their superior intelligence, they tried to rationalize their social instincts and came up with the idea of universal moral values, in which the things we do naturally are portrayed as having transcendental significance.

Darwin's views were considerably ahead of his time, and most nineteenth-century evolutionists preferred to believe that the human race was the predictable outcome of an inevitable ascent of the scale of mental and moral development. Darwin's heir apparent on the topic of mental evolution, George John Romanes, published his *Mental Evolution in Animals* (1883) and *Mental Evolution in Man* (1888) to identify the sequence by which successive levels of intellectual and moral activity were added to the human mind. His work was typical of a period in which it was taken for granted that identifying the course of development served as an explanation for why the development must have taken place—it was assumed to be inevitable that one stage of development would lead on to the next. For Romanes and most of his contemporaries, it was unthinkable that an unpredictable change of habitat and posture might have served as a key breakthrough without which the human race would not have come into existence. They assumed that the development of the brain (and hence of the mental faculties) was the driving force of human evolution. Standing upright and making tools were the consequences, not the causes, of increased intelligence. Increased socialization was also seen as a consequence of increased intelligence, and this was believed to have generated the superior levels of moral behavior in humans.

Some of Darwin's most important insights were thus ignored as "Darwinism" swept to popularity on the bandwagon of enthusiasm for progressionism. The clearest expression of this enthusiasm for the idea of the inevitability of progress was the popularity of Herbert Spencer's evolutionary philosophy. Spencer came from a humble background and became the spokesman for the aspirations of the middle class, which believed it had risen by its own efforts. He had supported the idea of biological evolution before the *Origin of Species* was published, and his *Principles of Biology* (1864) developed a detailed theory of biological progress. For Spencer, though, biological evolution was merely an inevitable implication of a cosmic philosophy of progress. He linked the biological origins of mankind with our mental, moral, and social progress into a seamless pattern of evolution, which was governed by universal forces. Spencer believed that the technological and cultural progress postulated by archaeologists and anthropologists had been made possible by a parallel development in the level of intelligence in the human race. Social progress was merely a continuation of biological evolution.

Yet Spencer's apparently materialistic worldview served only to modernize traditional moral values, not to destroy them. For him, the driving force of evolution was the individual's struggle to cope with the environment. This stimulated self-development, and the cumulative effect of generations of self-improvement constituted the progressive evolution of the species. The values that bought success in the race toward progress turned out to be those of the old Protestant work ethic: thrift, industry, and initiative. Evolution was driven by those values, Spencer believed, and naturally tended to favor development of them in the most advanced species. In Spencer's philosophy, the values cherished by the highest human civilization (industrial capitalism, of course) were the inevitable products of evolution because they expressed universal truths about what makes evolution work.

Most historians now accept that the conversion of the scientific world to Darwinism was made possible, at least in part, by the ease with which Darwin's theory could be made to seem congruent with Spencer's philosophy. As members of the rising middle class demanded more power within a rapidly industrializing society, they sought philosophies that would legitimize their demands for social progress by reform. The characteristics that brought success in commercial or professional life could be portrayed as moral values, allowing social progress to be seen as morally significant. Spencer's philosophy merely extended this idea of moral progress to a universal stage, allowing the traditional Protestant values to be seen as the driving forces of natural evolution. If Spencer himself did not speak openly of a God who created such a universe, his readers could see for themselves how his system allowed them to retain the values that had once been legitimized by religion. Liberal Protestants could thus accept many aspects of the evolutionary worldview without feeling that they were betraying their religious principles. As the scientific branch of Spencerian philosophy, Darwinism was inevitably seen as an expression of those same values.

Darwinism succeeded, therefore, because the more radical implications of natural selection and the branching model of evolution were subverted by the enthusiasm for progress. For anthropologists and archaeologists such as Lubbock, cultural progress could be depicted as a ladder, not as a tree: the various modern cultures were not different expressions of mankind's cultural diversity—they were merely stages in a single evolutionary sequence that all peoples must pass through. Modern primitives were merely relics of the past, preserved by isolation in areas where the climate provided no stimulus for economic development. Spencer's evolutionary view of sociology also stressed that industrial civilization was the inevitable goal of progress. To the extent that this linear and essentially goal-directed model of progress dominated late nineteenth-century thought, Darwin's emphasis on the divergent and open-ended nature of evolution went unheeded.

In the last decade of the nineteenth century, T. H. Huxley—once an architect of the original form of Darwinism—began to recognize some of the dangers inherent in the Spencerian approach. In his *Evolution and Ethics* of 1893 he attacked Spencer's belief that unrestrained free enterprise was the only route to progress. Huxley was now disturbed by the moral implications of a philosophy that left nature's failures to suffer and die. He wanted to preserve a traditional value system in which it was good to help your neighbor when he or she was in distress. To attack the logic of Spencer's argument, he now adopted a far more Darwinian view of the evolutionary process. For Spencer, free enterprise, with all its apparent harshness, was good because it was natural—and nature was inevitably progressing towards perfection. Any interference with nature would upset the trend and threaten progress. Huxley now argued that biological evolution was not, in fact, progressive—the fossil record showed that most developments were made to increase adaptive specialization, not to advance toward any absolutely higher state. If nature was not progressive, Huxley asked, why should we adopt its harsh methods in human society?

Huxley presented a very pessimistic philosophy in which humankind was trapped in a totally amoral universe, which threatened the most cherished human values at every point. He held out no hope of civilization surviving on a long-term basis in so hostile an environment—although he saw it as essential that we should all strive to maintain our human values in the face of everything that nature threw at us. This was certainly a very Darwinian view of the world, in the modern sense of the term, but it went largely unheeded. Even those who feared a degeneration of society as a consequence of the corruption of civilization nevertheless hoped that the progressive trend would be reestablished as soon as the right social conditions were created. Huxley's arguments show that even some of Spencer's original supporters now doubted that his philosophy was a truly moral one, but Huxley's pessimism was too extreme for his contemporaries to take him seriously. In any case, he had no way of explaining how the moral values that were such an important part of our humanity had ever managed to appear. In effect, he had left himself open to the charge that only a supernatural event could have produced creatures with values so diametrically opposed to those of nature itself.

In the twentieth century the more disturbing aspects of Darwin's theory were explored consistently, and by then there were many other factors at work undermining the Victorians' optimistic faith in progress. It was not until the 1930s and 1940s that the Darwinian theory reemerged as a powerful force in biology, following the "Modern Synthesis" with genetics. Some architects of the synthesis still hoped that evolutionism could be seen as a process with a moral purpose. Julian Huxley believed that evolution tended, in the long run, toward progress, in the sense that it generated animals that were more intelligent and less dependent on their environment. From this perspective, the human race could still be seen as the goal of evolution—and the theory gave its followers a sense of responsibility toward the rest of the natural world by confirming that humans are an integral part of the system. (Modern environmentalists may welcome the theory for the same reason.) Huxley even endorsed the theistic evolutionism expressed in Pierre Teilhard de Chardin's *Phenomenon of Man* (1959), which presented the human mind as the intended goal of a divinely created natural world. The popularity of Teilhard de Chardin's writings showed that the theological concerns underlying anti-Darwinian theistic evolutionism were still active in the mid-twentieth century.

Huxley seemed able to reconcile his Darwinian principles with Teilhard de Chardin's vague assumptions about a divine plan in evolution, but to many other biologists this kind of theistic evolution was a betrayal of science itself. The paleontologist George Gaylord Simpson stressed the materialistic implications of the new Darwinism and insisted that there was no sign of a moral purpose in evolution. Progress was an illusion, he said, because it was virtually impossible to define "progress" meaningfully in biological terms. Even by the crude definitions used by progressionists such as Huxley, most evolution did not, in fact, constitute an advance. For Simpson, the human race was merely the end product of a series of essentially unpredictable natural events. He emphasized the unrepeatability of evolution—a topic taken up in recent years by another paleontologist, Stephen Jay Gould, in books such as his *Wonderful Life* (1989). If the human race, as Simpson saw it, is a most unlikely product of the universe's activity, it becomes difficult to see how human moral values can be reflected in the laws of nature.

The very fact that recognition of these harsh realities (as Simpson would have called them) was so long delayed proves that evolutionism had not produced its initial impact on Western culture by destroying traditional values. The supposed clash between science and religion turns out to be more of a negotiation in which traditional values were modernized in accordance with the wishes of the now dominant middle classes. By the time biologists like Simpson began to emphasize the materialism of a Darwinian worldview, they no longer had the power to dictate social values. In reality, Darwinism as it came to be accepted in the twentieth century is but one of many cultural developments that have rendered the idea of a morally purposeful universe unpalatable to most intellectuals.

Religious fundamentalists in the United States seized upon the implications of the new Darwinism as part of their campaign to defend traditional Christianity, but their real concern—then as now—has been to return to a society in which the Bible has regained its old authority. The fundamentalist reaction against evolutionism began in the 1920s, when many rural Americans believed that urban society was becoming corrupt and sought scapegoats in various modernizing, intellectual trends. Darwinism was an obvious target because its apparent philosophy of trial and error in nature was so clearly opposed to the traditional idea of divine creation. Since the 1950s, however, most creationists have endorsed a view of the earth's history that rejects not only evolutionism but also the whole worldview of modern geology. They believe that the earth was created only a few thousand years ago, with the sedimentary rocks being laid down during Noah's flood. The selection theory is still a convenient target, but in principle the whole theory of evolution vanishes as a mere by-product of the creationists' alternative view of the earth's past.

Where, then, was the materialistic Darwinian bogeyman invoked by such conservative theologians and opponents of selectionism as Samuel Butler in the nineteenth century? Modern critics of Darwinism can point to Simpson and other biologists to confirm their claim that the theory destroys any sense of purpose in nature or in human life. But if Spencerianism—not Darwinism—was the dominant evolutionary philosophy of the late nineteenth century, historians need to reassess many of the traditional myths about the impact of Darwinism. It can no longer be asserted that Darwin's *Origin of Species* ushered in an age of rampant materialism in which all religious and moral values were abrogated. If Western civilization lost its traditional moral foundations, it did so in the early twentieth century, long after the Darwinian debate had died down. The loss of confidence resulted from the destruction of the optimistic progressionism of the Victorian age, which had temporarily allowed the old values to be retained in a modernized form. There were many early twentieth-century intellectual movements that contributed to this loss of faith, including Freudian psychology. But perhaps even more influential was the impact of World War I, which convinced many that Western culture's assumed air of superiority was unjustified. As we turn to the social implications of Darwinism, in chapter 4, we must be very careful not to fall into the trap of thinking that late nineteenth-century "social Darwinism" was based on a total rejection of traditional moral values.

Another problem of interpretation is posed by the character of Spencer's evolutionism. For opponents of Darwinism, such as Butler, the selection theory represented the worst aspects of materialism, while Lamarckism offered the prospect of a more humane evolutionary philosophy. Yet Spencer's own system was far more Lamarckian than Darwinian, since it emphasized the inherited effects of self-improvement in response to environmental challenge. One must recognize that the ideological labels pinned on theories by Butler and other writers are products of their own cultural circumstances, and are therefore not absolutely valid indications of the theory's moral standing. If Spencer's ruthless philosophy of progress through success was a manifestation of the new evolutionism's materialism, that materialism was nevertheless linked to a theory that was Lamarckian rather than Darwinian in the modern sense of the term. The allegedly more humane, non-Darwinian mechanism quite easily became the basis for social policies that later critics would dismiss as the worst kind of "social Darwinism."

To see Darwinism as a scapegoat for the ills of twentieth-century culture is to ignore the character of late nineteenth-century evolutionism. It is all too easy to project a modern understanding of Darwin's theory back onto the earlier debates, assuming that what disturbs its modern critics was apparent at the time it was introduced. Conservatives of Darwin's day may have correctly assessed the dangerous implications of the selection theory from the start, but their objections were ignored because the first-generation evolutionists were only able to take up the theory by assimilating it into a philosophy of universal progress. By the time the true implications of Darwinism had been explored by twentieth-century biologists, science had lost its ability to dominate our culture. Once we recognize that nineteenth-century Darwinism was not the same as its modern counterpart, we are forced to reassess both the cultural impact of evolutionism in general and the ideological labels that various writers have attached to the different evolutionary theories.

# 4

# Social Darwinism

The fact that evolution theories had social implications is taken for granted by many scholars dealing with late nineteenth-century thought. The term "social Darwinism" has come into widespread use to denote the application of Darwinism in this area. But in recent years the use of this term has become problematic because it has been used as a label for a particular kind of interaction between biology and sociology, an interaction that is more the product of hindsight than of careful historical analysis. The early form of "Darwinism" was not presented as a purely materialistic philosophy that destroyed all moral values, and when we examine various kinds of "social Darwinism" more closely, we can see that in many cases the biological theories used as the basis for social arguments were actually non-Darwinian in origin. This does not imply that we should sever the presumed link between biological and social theory, but we must accept the fact that a wide range of biological theories was available for use by those who wanted to claim that their preferred value system was more "natural" than their rivals'.

The classic expression of the view that Darwin's theory had a harmful impact on nineteenth-century social thought is found in Richard Hofstadter's *Social Darwinism in American Thought*.[1] According to this interpretation, Darwin's evolutionary mechanism, the "survival of the fittest" in the "struggle for existence," was used by Herbert Spencer and his followers to legitimize the ideology of laissez-faire individualism that sustained the competitive ethos of capitalism. The free-enterprise system would produce economic and social progress because those individuals unable to cope with the new industrialized life-style would be killed off by starvation. It was only natural that useless members of society should be eliminated to leave room for their more efficient fellows. Later on, the same policy was applied to justify the domination and extermination of "inferior" races in the age of imperialism. Darwinism is thus seen as having ushered in an era in which the motto Might Is Right governed human relationships and traditional moral values were consigned to the garbage can of history.

Commentators who see Darwinism as the foundation stone of a materialistic worldview, one in which all moral values are abrogated in the name of expediency and profit, assume that these policies can be attributed to the influence of the theory of natural selection. They see Darwinism as an amoral theory because it destroys the meaning of human life and encourages us all to behave like animals. By contrast, rival theories such as Lamarckism have been presented as vehicles for a more humane message. Hofstadter used the work of sociologist Lester Frank Ward as an example of how Lamarckism could be used to sustain a policy of state intervention to direct society (and the human race) toward moral improvement via social reform. Ward argued that if Lamarckist theories were correct, the improvements in individual health and behavior brought about by reform would become imprinted on future generations, giving a permanent boost to the human race's evolution. Modern creationists also attribute the decline of moral values to Darwinism (although some American creationists are among the most extreme advocates of unrestrained free-enterprise).

The possibility that this negative image of nineteenth-century social Darwinism was an artifact of the historians' imagination was raised by Robert Bannister. According to Bannister, Hofstadter's interpretation was a myth created by exaggerating the amoral character of Spencer's beliefs and those of his followers. The purpose of this myth was to legitimize the reformist policies of the early twentieth century by contrasting them with the harshness of what had gone on before.[2] Bannister implied that Spencer's philosophy was not, in fact, very Darwinian; critics argued that he was trying to defend the objectivity of science by driving a wedge between biological and social Darwinism. There has certainly been great debate over the extent to which Darwin's theory of natural selection reflected the values of Victorian capitalism.[3] Some scientists have tried to show that Darwin's theory was conceived and popularized on the basis of objective evidence, with no input from its social and cultural environment. But few modern historians would argue that scientific theories are conceived in a vacuum, nor does it seem likely that a theory as controversial as Darwin's could come to dominate cultural life merely because it had good evidence to back it up.

The critical issue is the nature of the interaction between science and social values. The fact that Darwinism had a social dimension does not mean that we can impose modern definitions of Darwinism (or of any associated ideology) on the past. Bannister's point was that a social Darwinism based on the worship of brute force was never as popular as Hofstadter had claimed. What has been perceived as a totally amoral philosophy of struggle was presented at the time as a vehicle for transforming, not destroying, moral values. This view has been sustained by later studies confirming that the term "social Darwinism" was used only in a perjorative sense, and did not, in any case, come into use until the early twentieth century.[4] There were few card-carrying social Darwinists in the late nineteenth century because no one—not even the scientists—saw natural selection as an adequate mechanism of biological or social evolution. "Darwinism" was certainly linked to Herbert Spencer's philosophy of progress, but this was neither totally amoral, nor very Darwinian (in the modern sense of the term). To explore the links between evolutionary biology and social thought, historians must take into account the complexity of the debate over the nature of the evolutionary process.

When one turns to later varieties of social Darwinism, the problem of definition becomes even more acute. Most late nineteenth-century anthropologists and biologists assumed that the dark-skinned races of humankind were inferior to the white, and it was often thought inevitable that where the two came into contact, the dark race would be eliminated. Since there was talk of the survival of the fittest races, these attitudes are easily portrayed as a later form of social Darwinism. Yet the concept of a racial hierarchy was so pervasive that it was endorsed by the supporters of all the rival evolutionary theories, as well, especially by Lamarckians such as Ernst Haeckel and Edward Drinker Cope. Darwinians certainly wrote about the elimination of unfit races—but so did anti-Darwinians, because they saw natural selection as a purely negative effect that merely got rid of the less successful forms produced by some more purposeful evolutionary process. As we saw in chapter 3, this negative side of selection was frequently endorsed by evolutionists who postulated non-Darwinian mechanisms to explain the origin of the fitter characteristics.

Ideas of racial conquest and extermination were widespread and had been used throughout the nineteenth century by conservative writers who opposed the liberal progressionism that characterized Spencer's social philosophy.[5] In an age of imperialism, it seemed only natural to revive models of social development based on the rise and fall of empires in the course of human history. It could be argued that Darwinian biogeography, which explained the distribution of species in terms of migrations and replacements, was modeled on this conservative or imperialistic ideology. Yet both Darwinians and anti-Darwinians routinely borrowed the metaphors of conquest when describing the history of life on earth—and the relationships between human races.

The claim that "social Darwinism" was not very Darwinian does not imply that there was a rigid dividing line between science and ideology. Nineteenth-century evolution theory was very different from its modern counterpart. To say that Spencer and his followers adopted a social policy based on the survival of the fittest among individuals subject to random genetic variation would be to make them "social Darwinists" in a modern sense. But it is now clear that very few biological evolutionists—let alone social evolutionists—were in a position to take such a suggestion seriously in the nineteenth century. To understand the social *evolutionism* of that period, we need to forget modern Darwinism and relate the various social policies to the scientific theories of the time. If we accept that the early form of "Darwinism" had components that were non-Darwinian by modern standards, we can better appreciate the true character of Spencer's efforts to produce a universal philosophy of evolution. At the same time, we shall be able to see how some anti-Darwinian theories could be used to justify social policies that might easily be mistaken as "Darwinian" by a reader who was not aware of the complexity of the biological sciences at the time.

The character of both the prevalent social values and the evolution theories have changed significantly since the mid-nineteenth century. To claim a direct link between the publication of Darwin's theory in 1859 and the imperialist view of race relations in the early twentieth century is absurd. If a connection is sought, it must take into account the intervening developments in biology as well as the changing character of imperialist attitudes. To interpret all social arguments derived from biology as expressions of Darwinism is equally absurd, since we now know that many different evolution theories were under consideration. All biological theories, Darwinian and non-Darwinian alike, could be used to justify social policies that most of us now regard as reprehensible.

These historical reconsiderations inevitably interact with modern attitudes. Scientific theories acquire labels that are supposed to denote their wider implications, and those who find these labels a convenient form of shorthand are often annoyed by efforts to show that the labeling process has involved an oversimplification of history. Left-wing commentators have found the term "social Darwinism" useful for identifying all social policies in which nature is allowed to take its course, with the apparently inevitable implication that the least able members of society will be left to suffer and die. These commentators are sometimes less than willing to acknowledge the fact that Spencer's justification of laissez-faire policy was based more on the Lamarckian than on the Darwinian component of his biology. Lamarckism has frequently been portrayed as a theory that is intrinsically less materialistic, and hence less threatening, than Darwinism. Anti-Darwinian writers such as Arthur Koestler have ignored the evidence suggesting that Lamarckism, too, played a role in promoting unsavory attitudes such as racism. Any analysis of social evolutionism must acknowledge the existence of both Darwinian and non-Darwinian efforts to show that intolerant social policies are inescapable expressions of human nature.

The historical reconsideration of Darwinism does not seek to sever the link between science and social values, but rather to show that the relationship is so complex that the traditional labels attached to theories are dangerously simplistic. Dangerous, because people can easily be misled into thinking that a certain theory offers a guarantee of ideological purity, little realizing that a slight twist in the arguments can turn the same theory into a recipe for what they would regard as a social nightmare. History tells us that scientists and their followers have been able to use all sorts of different theories to uphold the same social policy—and that different policies can be justified on the basis of the same theory. Some of the links gain more publicity than others, and these become the "labels" that later writers use to identify the ideological function of the theories. But the labels are socially constructed in particular circumstances—they do not correspond to real relationships of the kind that would justify the assertion that a certain theory is the most obvious or most natural way to project a certain value-system onto nature. Some links may look more convincing than others in retrospect, but to ignore the historical context in which the links were forged is to create an oversimplified picture of the way in which theories and social policies are related.

## Social Evolutionism

The reinterpretation of the history of evolutionism offered above allows us to explore the complexities of the relationship between nineteenth-century models of biological and social progress. The idea that social Darwinism is an amoral philosophy based on the rejection of all traditional values is based on the identification of Darwinism with the image of ruthless struggle. It assumes that if one advocates leaving everything in society to the self-interest of individuals, then there can be no room for the expectation that social development has a moral goal. All individuals are struggling against one another in the mere hope of survival, caring nothing for what happens to anyone else or to society as a whole. If progress results from this abrogation of all concern for others, it can only be *economic* progress—the generation of more material goods to satisfy mere physical appetites.

This emphasis on the theme of struggle is popular among critics who maintain that the free-enterprise system is based solely on self-interest, that those with wealth or ability seek only to enjoy the fruits of their good fortune without caring for those who are less well endowed. Such a model of the free-enterprise society certainly seems "Darwinian" in the sense that good and ill fortune are random, and those who benefit have no interest in modifying the natural outcome of the process by which the unfit are cast aside. This may well be a valid assessment of capitalist behavior, but it is not an accurate picture of the arguments that were used to defend the system. Darwinism gained a foothold in nineteenth-century thought because it emphasized the progressive nature of evolutionary change. Progress was still defined in moral rather than purely material terms, and it is a mistake to see Spencer and his followers as advocates of pure materialism.[6] The whole point of Spencer's philosophy was that nature should be allowed to progress without interference, because the ultimate goal of social development was a moral one. We may suspect that Spencer's arguments were merely words designed to conceal the true selfishness of the commercial elite, but to understand the public debate at the time we have to take them at their face value.

If Darwin's theory became associated, in the public's mind at least, with Spencer's social philosophy, it was being identified with a model of progress through struggle that does not correspond to the modern interpretation of natural selection. In the strictly Darwinism mechanism, individuals are born with fit or unfit characteristics (as measured by the local environment) and can do nothing to influence their fate. Natural selection is a kind of genetic Russian roulette, and no effort on the part of the individual organism (or human being) can save it if it is born with harmful characteristics. But it was certainly not Spencer's purpose to invoke the concept of selection acting on a random distribution of characteristics that were rigidly determined by heredity. The whole purpose of the struggle that must take place within a free-enterprise society, as Spencer saw it, was not to eliminate the congenitally unfit, but to force everyone to become fitter. Liberal thinkers endorsed Samuel Smiles's vision of a society based on self help (*Self Help* was in fact the title of his book, published in the same year as *Origin of Species*). Spencer merely generalized this emphasis on self-improvement to make it the basis of all social progress. When a strictly hereditarian view of human nature did emerge at the end of the century, it was associated not with free-enterprise individualism, but with calls for the state to prevent the reproduction of "unfit" members of society.

Spencer argued that a policy of state aid for the less fortunate was dangerous because it protected everyone from the chief stimulus to self-development. If people could be sure of a living provided by the state, why would they work hard to develop their own potential? Only harsh necessity would force people to exercise their abilities to the fullest. Anyone who made a mistake would be taught a lesson by suffering the consequent penalties—and this would teach him or her to do better next time. Having thus learned how to function effectively in society, people would pass the message on to their children, thus perpetuating the effect of the lesson in future generations:

> If to be ignorant were as safe as to be wise, no one would become wise . . . . Unpitying as it looks, it is better to let the foolish man suffer the appointed penalty of his foolishness. For the pain—he must bear it as well as he can: for the expe-

rience—he must treasure it up, and act more rationally in the future. To others as well as to himself will his case be a warning. And by multiplication of such warnings, there cannot fail to be generated in all men a caution corresponding to the danger to be shunned.[7]

Spencer admitted that the policy seemed harsh, yet its ultimate purpose was to force everyone to become more intelligent and more adaptable. He believed that nature is the best teacher when it comes to promoting the Protestant virtues of thrift and industry.

Spencer did not rule out the possibility that, in the most extreme case, some unfortunates might actually die as a consequence of their inability to cope. He called this the indirect way of bringing the species into equilibrium with its environment. It was to describe this mechanism—which is, of course, natural selection—that Spencer used the phrase "survival of the fittest" in his *Principles of Biology*.[8] The fact that Spencer coined this notorious phrase has fueled claims that he was the original social Darwinist, yet for him this was always a secondary mechanism. More important was the direct mechanism of adjustment by which individuals survived and learned from their experience, thus improving themselves and transmitting the improvement to future generations.

This was an explicitly Lamarckian process: struggle was the spur to self-development, and the evolution of society (or the species) was the sum total of generations of such individually acquired characteristics. To any twentieth-century thinker experienced in the concepts of Mendelian genetics, there is a fatal flaw in the argument. The transmission of knowledge through learning is quite different from the transmission of biological characteristics through heredity. But in the period before modern genetics hardly anyone made this crucial distinction, and much support for Lamarckism derived from a failure to distinguish between the cultural and biological modes of transmission. Spencer and his fellow Lamarckians simply assumed that the transmission of learned characteristics would eventually spill over into the transmission of biologically implanted instincts. Spencer's reliance on Lamarckism is confirmed by the fact that he later defended the theory against the neo-Darwinians of the 1890s. His social philosophy was not an expression of rigid hereditarianism; on the contrary, it depended on individuals having the ability to improve themselves when suitably stimulated. Even though he has been called a social Darwinist,

Spencer was really a social Lamarckian. He justified a society based on struggle not by reference to natural selection, but by portraying struggle as the chief force encouraging self-development.

The Lamarckian component in Spencer's thought has been ignored because evolution theories have all too often been portrayed as rigidly polarized in their social implications. Lamarckism is instead identified with the social reformism of Lester Frank Ward and others, who sought to improve the human race by using education to foster the characteristics they wished to see enhanced in future generations. They, too, assumed that traits imprinted through education would eventually find their way into the biological constitution of the race. This was a form of social Lamarckism, but it was not the only form. Spencer dismissed claims that we can control the future development of society, because he thought that nature was too complex for us to be able to predict the results of our actions. Progress would only be ensured by letting nature take its course; individual competition, not a state-controlled school system, was the best way of teaching the human race to improve itself. Historians have ignored Spencer's alternative version of social Lamarckism because they have allowed themselves to be influenced by an artificially rigid characterization in which Darwin's selection theory is seen as the only way of projecting the capitalist ethos onto nature.

If Darwin's theory could become associated in the public mind with Spencer's philosophy of social progress, then it is probable that many ordinary people simply failed to see the distinction between the Darwinian and Lamarckian modes of progress through struggle. In the decades before the appearance of Weismann's concept of germ plasm (let alone the modern concept of the gene), biologists were so vague about the production of variations that no clear distinction between theories was possible. Darwin himself accepted that variations were caused by an environmental disturbance of the reproductive process, and he never ruled out the Lamarckian effect (that is, directed rather than undirected variation). Spencer's belief that competition would stimulate self-improvement would thus have seemed compatible with what many people saw as "Darwinian" evolution.

The struggle motif could also be developed in other, equally un-Darwinian, ways. Even Spencer himself realized that, in addition to the struggle between individuals, society as a whole could be said to be struggling against the limitations of the external environment. It was for this reason that Spencer and many Lamarckians believed that a harsh climate was important to social evolution, because it provided a greater stimulus or challenge to humans. One of the leading American advocates of what has been called "social Darwinism," William Graham Sumner, was in fact more concerned with the struggle of society as a whole against the limitations imposed by nature than he was with individual competition. Yet this level of struggle was by no means an exclusive component of the Darwinian theory—on the contrary, it was invoked quite easily by Spencer and other Lamarckians.

Biologists such as Ernst Haeckel also projected an image of evolution progressing to ever-higher states of development in response to the constant challenge of the environment. Here too the emphasis switched from the struggle of individuals against their fellows to the struggle of the whole species against the environment. The more stimulating the environment, the greater the challenge to individuals and the more rapid the rate of progress for the species. When E. Ray Lankester wrote about the threat of degeneration in 1880, he used biology to argue that the human race itself might degenerate if it conquered the environment so effectively that no challenge was left. By the end of the century there was widespread fear that progress had reached a peak and that the human race might soon degenerate. In 1895 the author H. G. Wells borrowed Lankester's ideas for his science-fiction story *The Time Machine,* which depicts a future in which the human race has degenerated through lack of stimulus.[9] Lankester and Wells claimed to be Darwinians, and it could certainly be argued that natural selection would be more active in a harsh environment. But it was also argued by many that the environment was a challenge directed primarily against the individual organisms, whose responses to it would accumulate to produce progress or degeneration in the species. Some exponents of this model of progress through struggle against the environment openly joined the Lamarckian camp during the debates at the end of the century, including the embryologist E. W. MacBride. MacBride thought the lower forms of animals had been left behind in the march of progress because they had adopted a less stimulating life-style.

## Racial Conflict

In the 1920s, E. W. MacBride defended the Lamarckian experiments of Paul Kammerer, who claimed to have demonstrated the inheritance of acquired characteristics in amphibians exposed to new environments. Arthur Koestler, in his account of this, *The Case of the Midwife Toad* (1971), refers to MacBride as "the Irishman with a heart of gold," apparently under the impression that all Lamarckians must be nice people. In fact, MacBride was an Ulster Protestant by birth who thought that the native Irish should be sterilized lest they undermine the biological character of the British race.[10] Like many Lamarckians, including the paleontologist E. D. Cope, MacBride was convinced that evolution had generated a hierarchy of human races in which the whites, and especially the Anglo-Saxons, were the most highly developed. He believed that those races that had evolved in a harsher and hence more stimulating environment would have developed higher levels of intelligence in response to the challenge. Cope depicted the African blacks as a race that retained many primitive characteristics. MacBride saw the Irish as a similarly primitive type which had originally evolved in the softer climate of southern Europe. Just as Lamarckism could be used to justify optimistic philosophies of social reform, it could be used equally well as the foundation for highly intolerant views—a point that Koestler and many later anti-Darwinians find difficult to appreciate.

The fact that Lamarckians were deeply involved in the effort to provide an allegedly scientific basis for racism highlights the problems that have been generated by some historians' branding of late nineteenth- and early twentieth-century imperialism as a form of "social Darwinism." Certainly there were efforts made to justify both the conquest of "inferior" races and the increasingly belligerent attitude of the European powers toward one another in terms of a "struggle for existence" between groups rather than individuals. Some scientific Darwinists were involved in this—but so were many biologists whose views were profoundly non-Darwinian in nature. Here again, the concept of struggle was applied so loosely that it could be used by scientists and social thinkers with a wide range of opinions.

Anthropologists in the late nineteenth century developed a theory of cultural evolution in which it was assumed that societies naturally evolved from the simplest hunter-gatherer stage toward industrial capitalism. This concept of a fixed ladder of developmental stages was promoted in Britain by Edward B. Tylor and in America by Lewis Henry Morgan. Archaeologists such as John Lubbock emphasized that modern "savages" were the equivalents of our Stone Age ancestors, frozen in a prehistoric level of development. It was widely assumed that the level to which any particular society had ascended depended on the degree of environmental stimulus. The white races of Europe were believed to be more highly developed than their black or brown counterparts because their ancestors had been exposed to a harsh northern climate rather than an enervating tropical one. The most technologically primitive modern societies (the Australian aborigines were a favorite example) were relics of the distant past, preserved in the present because they had failed to advance any further. It was easy to see this kind of cultural evolution as a continuation of the biological process by which life had advanced toward the human level of intelligence. At all levels, the anthropologists thought, progress toward higher levels of organization occurred because individuals responded to an environmental challenge by developing themselves, and then passed their acquired characters on to later generations.

Tylor had originally developed his theory of cultural evolution to preserve the traditional belief that human nature was everywhere the same. Societies were different not because people were different, but because people with the same mental equipment had risen to different levels of cultural development in response to different levels of external stimulus. But Spencer's insistence that social and biological evolution were aspects of the same process threatened to destroy this sense of potential human unity. According to Spencer, social and mental evolution went hand in hand: intelligent people developed a more complex culture, and a complex culture in turn stimulated greater self-development. After millions of years of exposure to a different intensity of stimulus, two races would have acquired not only different levels of culture but also different levels of mental ability. Just as lower species were stuck at inferior levels of biological development, so lower races were stuck at inferior levels of mental evolution. The theory of evolution through cumulative self-development led, in the end, to racial determinism. It was believed that some races were so far behind that, whatever the stimulus, they could never catch up with the triumphant whites.

The optimistic side of the philosophy of self-development was thus reserved for the whites. It might apply in theory to other races, but their potential for improvement was restricted by the inherited effect of generations of exposure to a less stimulating environment. As the nineteenth century drew to a close, this view of racial character was increasingly used to justify the growing tide of imperialism, and the theory of biological evolution played a vital role in legitimizing this development. Once it was accepted that humans had evolved from the apes, it was difficult to resist the implication that some branches of the human tree might have grown faster and higher than others. In the *Descent of Man* Darwin took the racial hierarchy for granted, citing the conventional view that the whites had a larger cranial capacity than other races.

In this way the evolutionists became convinced that, despite their common origin, some races were more advanced than others. This immediately raised the question of what would happen when two races, originally separated by a geographical barrier, came into contact. Since the "lower" race would not be able to raise itself sufficiently by self-development, the mental differential would persist, and it was only a short step from this idea to visualizing the interaction in terms of the struggle for existence. Darwin certainly did not advocate the deliberate extermination of so-called lower races, but he assumed that, for complex reasons, such races gradually declined in numbers whenever confronted by invading whites. By the end of the century the Darwinian biologist Karl Pearson was openly advocating the replacement of inferior races by white populations wherever this was possible: "It is a false view of human solidarity, a weak humanitarianism, not a true humanism, which regrets that a capable and stalwart race of white men should replace a dark-skinned tribe which can neither utilise its land for the full benefit of mankind, nor contribute its quota to the common stock of human knowledge. The struggle of civilised man against uncivilised man and against nature produces a certain partial 'solidarity of humanity' which involves a prohibition against an individual community wasting the resources of mankind."[11] In a footnote to this passage, Pearson warned against the brutality of simple extermination, but nevertheless rejoiced that whites were replacing the natives of America and Australia.

Pearson was also an imperialist, and he urged the British to defend their empire against the threat from rival powers such as Germany. When the recruitment drive for the British army in the Boer War revealed that many working-class men were unfit to serve, Pearson wrote his *National Life from the Standpoint of Science* (1901) to raise the specter of racial degeneration threatening to undermine the nation's position in the international power struggle. But Pearson was no Lamarckian—he followed Francis Galton in assuming that the character of each individual was rigidly determined by inheritance. He thus argued that to protect the nation's interests, the "unfit" members of society should be prevented from breeding. Only the fittest—the professional classes and skilled workers—should be allowed to transmit their superior traits to future generations. The "eugenics" movement (the term was coined by Galton) argued for a state-controlled policy of artificial selection for the human race. This was a truly hereditarian social philosophy, in which the individual had no hope of improving his or her inherited character. No amount of social reform could improve the biological character of those born unfit—but centralized, state-run, artificial selection would have to take the place of elimination by natural causes.

It would be easy, once again, to see eugenics as a form of social Darwinism—but it was based on the very opposite view of the state to that upheld by Spencer and the advocates of laissez-faire. Nor was the eugenics movement supported only by Darwinians. Pearson was a genuine Darwinian in his biology, a staunch defender of the selection theory at a time when it was challenged by many biologists. Yet MacBride's call for the Irish component of the British population to be sterilized was an extreme version of a eugenics policy, and he was a Lamarckian. Some early geneticists also supported eugenics, although they rejected the Darwinian selection theory as an explanation of natural evolution. In America, for example, C. B. Davenport tried to argue that unfit characteristics such as feeblemindedness were transmitted by single genes, which might be eliminated from the population if the affected individuals were prevented from breeding. Darwinism thus provided only one of several scientific foundations for the movement.

Various kinds of evolutionism contributed to the ideology of the race hierarchy, while Darwinism supplied the rhetoric of struggle that could easily be translated from the individual to the racial level. The image of nations competing in a "struggle for existence" to determine which should rule the world became popular in the decades running up to World War I. We have already noted this in the writings of Karl Pearson, and it was widely believed that the same way of thinking contributed to the ruthlessness of the German war machine. This was certainly the view of the American biologist

Vernon Kellogg, who visited the invading Germans in Belgium and wrote his book *Headquarters Nights* to warn against this aggressive version of social Darwinism.[12] Following the war, Germany had a strong eugenics program long before the Nazis came to power; the Nazis extended it in their efforts to "purify" the Aryan race so that it could play its destined role in the global struggle.

That Darwinism played a role in stimulating this atmosphere of national and racial rivalry cannot be denied. But at the same time it is important to note that there were other beliefs in biology, anthropology, and archaeology that could also have played a similar role, some of them deriving from anti-Darwinian sources. In the early nineteenth century, German historians under the influence of an idealist philosophy had developed a model of human progress based on principles very different from those of the liberal tradition embodied in Spencer's philosophy. Where the liberals saw progress as the cumulative effect of individual efforts over many generations, the idealists presented history as a sequence of distinct episodes, each representing a new development of the human spirit, and each produced by a new nation or race. This was extended back into prehistory by philologists such as Max Müller, who postulated that waves of Aryan invaders had streamed into Europe from a mysterious center of progress in the east. (Philology is the study of languages; Aryan is the root from which later European languages, including Greek, Latin, and the Germanic languages, evolved.) Many archaeologists believed that the population of Europe was made up of a mosaic of peoples, each the remnant of a different race that had invaded the continent and then been marginalized by later waves of immigrants.

By focusing on successive invasions, this approach to prehistory necessarily interpreted the relationship between different peoples in terms of conflict, but it did so quite independently of Darwinism. Biologists studying the geographical distribution of animals and plants also had to invoke an explanatory scheme based on the migration of species to new territories and their extermination of earlier species. Darwin himself certainly used this model, with all its imperialist metaphors, but so did biogeographers who had no interest in his theory of natural selection as the driving force of adaptation. In the early twentieth century, the fossil record was widely interpreted in terms of the rise and fall of dominant groups such as the dinosaurs. Thus in the study of both animal and human prehistory, conservative or idealist models of development interacted with the liberal concept of gradual evolution. The liberal or Darwinian-Spencerian approach had dominated the study of human origins during the early phase of the evolution debate, but in the age of imperialism at the end of the century, the rival model was revived as more biologists and anthropologists became concerned with racial distinctions and with the possibility that conflict between races is a major factor in human evolution.

In his book *Ancient Hunters and Their Modern Representatives* (1911), the geologist and archaeologist W. J. Sollas clearly recognized that the racial-conflict model of prehistory had broader implications. He interpreted the various marginalized peoples of the world as relics of earlier stages in human prehistory, but emphasized that their isolated position was the result of their having been driven from the earth's main landmasses by more successful later races. He believed that the principle of "might is right" was nature's way of ensuring the success of more highly evolved types:

> What part is to be assigned to justice in the government of human affairs? So far as the facts are clear they teach in no equivocal terms that there is no right which is not founded on might. Justice belongs to the strong, and has been meted out to each race according to its strength; each has received as much justice as it deserves. What perhaps is most impressive in each of the cases we have discussed is this, that the dispossession by a new-comer of a race already in occupation of the soil has marked an upward step in the intellectual progress of mankind. It is not priority of occupation, but the power to utilise, which establishes a claim to the land. Hence it is a duty which every race owes to itself, and to the human family as well, to cultivate by every possible means its own strength: directly it falls behind in the regard it pays to this duty, whether in art or science, in breeding or in organization for self-defence, it incurs a penalty which Natural Selection, the stern but beneficent tyrant of the organic world, will assuredly exact, and that speedily, to the full.[13]

The imperialist message of this passage is obvious enough, and it would be tempting to dismiss Sollas as the worst kind of "social Darwinist," obsessed with natural selection as the vehicle of progress. His words seem to parallel those of Karl Pearson. Yet elsewhere in his book Sollas ridicules the Darwinian mechanism as being impotent to explain the generation of higher characteristics. Like many other critics of Darwinism, Sollas was prepared to accept natural selection as a mechanism for weeding out the types left behind in nature's progress—but he confessed himself unable to explain how the higher types were, in fact, evolved.

At the same time, many archaeologists were becoming convinced that the early human type known as Neanderthal man was—despite its ape-like appearance—not ancestral to modern humans. The original Neanderthal remains had been discovered in 1857 and had been widely discussed in the evolution debate because this very ancient human type had a skull with heavy brow ridges like an ape's. Under the influence of the model of continuous, linear progress, the first generation of evolutionists had interpreted the Neanderthals as a stage in the development from the apes leading toward the lowest still-living races. But now anatomists such as Arthur Keith began to insist that the Neanderthals were so different that they could not be the ancestors of modern humans. They were instead believed to be a branch of the human family tree that had been wiped out by our own ancestors at the end of the Mousterian period of the Palaeolithic or Old Stone Age (Lubbock's term for the age of primitive chipped-stone tools). The notorious Piltdown fraud of 1912, in which spurious human fossils were planted to create the impression that there was a new kind of Stone Age human, was taken seriously at first precisely because it seemed to substantiate the idea of a non-Neanderthal line of human evolution.[14]

Here the model of discontinuous evolution based on racial conflict was incorporated into the heart of the scientific account of human origins. Keith—like Sollas—clearly saw the theory of Neanderthal extermination as a justification for imperialism. Keith had become acutely conscious of the racial differences among modern humans, and he favored the view that racial interaction led to struggle and extinction. Racial and national competition was necessary to sustain progress in the modern world, Keith believed, but this was merely a continuation of the process that had been going on since the Neanderthals were wiped out: "What happened at the end of the Mousterian period we can only guess, but those who observe the fate of the aboriginal races of America and Australia will have no difficulty in accounting for the disappearance of *Homo neanderthalensis*. A more virile form extinguished him."[15] In later years Keith went on to develop a whole theory of human evolution based on racial competition. Even so, he was never a supporter of the Darwinian theory of how new races or species were produced. Competition was only a secondary factor, Keith believed, weeding out those races that did not advance fast enough.

The opinions of Keith and Sollas reveal that an emphasis on racial struggle need not have its origin in the Darwinian theory. Some Darwinians certainly did translate the struggle for existence from the individual to the racial level, and no doubt everyone made use of the rhetoric of "struggle" and "survival" encouraged by Darwinism. But the idea that races must come into conflict to determine their destiny had been used by historians and archaeologists before Darwin published, and it was used into the twentieth century by scientists and other writers whose ideas have their roots in distinctly non-Darwinian models of development. Darwinism was a theory of gradual, continuous change, and these rival ideas were based on a theory in which progress took place through a series of distinct steps in which more highly developed types suddenly began to impose themselves on the world. The Darwinian-Spencerian model of continuous development was favored by liberals who saw progress as the cumulative effect of individual efforts over many generations. But this theory had always been opposed by a more conservative ideology in which the character of races and nations was thought to be determined by forces that transcend the life of the individual. To attribute the imperialist model of racial conquest solely to the influence of Darwinism is to conceal the ongoing influence of other historical models, which derived from a more conservative ideology that had always stood opposed to the liberal progressionism of the Darwinian-Spencerian tradition.

## The Reaction against Evolutionism

In the early twentieth century, the social sciences were able to establish themselves as separate disciplines, in part by asserting their independence from the nineteenth-century evolutionary paradigm. Social and cultural evolution were to be treated as processes that needed to be studied independently of any model provided by biology. In America, cultural anthropologists led by Franz Boas rejected the linear pattern of cultural evolution in which modern Western values were assumed to be the high point of development. They insisted that each culture had to be understood in its own context—that it was illegitimate to evaluate one culture by the standards of another. The cultural relativism implied by this approach struck at the heart of the assumed superiority of industrial capitalism. The anthropologists and sociologists did not deny that the human race had evolved from an animal ancestor, but they were not prepared to allow this ancestry to dictate the behavior of modern humans. Biology might have given us the capacity for culture, but that capacity could be expressed in many different ways.

Significantly, Boas and the other anthropologists thought they were rejecting "Darwinism," although what they were actually turning their backs on was the nineteenth-century evolutionary model that was only partly inspired by Darwin's theory. Indeed, there is a sense in which the cultural relativism of the new approach paralleled very closely the Darwinian view of evolution as a branching, divergent process that had no main line of development and no predictable goal. Natural selection, however, had no role to play in the anthropologists' view of cultural history. They believed that people occasionally form new habits and transmit them by teaching them to the next generation, a process that is more akin to Lamarckism, except that the "inheritance" is cultural rather than genetic.

Boas and his colleagues were convinced that most of the differences in mental and moral character attributed to the various human races were cultural rather than biological in origin. They thus campaigned vigorously against racist biology and against the eugenics movement. During the early decades of the twentieth century, the biologists fought back, insisting that the anthropologists and sociologists were ignoring fundamental human characteristics—and character differences—implanted in us by the evolutionary process. The debate took on moral and political overtones, with the social scientists insisting that the human race could be improved by reforming the ways in which people were encouraged to interact with one another. According to their view, one way of behaving was not more "natural" than another, and so a better society was possible. Biologists insisted that this approach ignored built-in human instincts, which compel people to act in certain ways and which it is folly to ignore in any proposed social restructuring. This new form of biological determinism has often been called social Darwinism, although its scientific foundations may actually lie in genetics or in other areas of biology.

The ethical side of the social-biological debate, which continues today, conceals significant professional interests: it was and still is to the social scientists' advantage to claim that they have the key to understanding society, since this allows them to demand government funding for their research and its applications. By resisting this trend, biologists can assert that their expertise is also relevant and can put in counterbids for the same funding. Professional rivalry may underline much of the high-sounding rhetoric of the debate over the ethics of "social Darwinism."

As the social scientists emancipated themselves from evolutionism, biology itself was undergoing a revolution in which Darwinism was synthesized with genetics. Far from encouraging the biologists to extend their case for a deterministic view of human nature, however, the emergence of the modern Darwinian synthesis had precisely the opposite effect. According to Julian Huxley and other architects of the modern synthetic theory of evolution, the genetic structure of the human species confirmed that there were no major evolutionary divergences between the races. From the 1940s onward, many scientists began to oppose the racism that had flourished during the era of non-Darwinian evolutionism. Nor did the new Darwinism endorse the idea of racial conflict as a mechanism of natural selection. Huxley and the evolutionists joined with the anthropologists who had opposed racism. The eugenics movement also fell into disrepute, as the new findings in biology made it seem unlikely that any policy of controlled breeding could significantly affect the level of "harmful" characteristics in the human race. Admittedly R. A. Fisher, one of the founders of population genetics, was a strong supporter of eugenics even though his mathematical formulas showed just how slow any selective effect would be. But other biologists, led by J. B. S. Haldane, rejected eugenics on both scientific and moral grounds.

In the end, scientific racism and eugenics were overthrown as much for political as for scientific reasons. The new Darwinian evolutionism certainly helped to undermine many of the oversimplified arguments used to defend biological determinism, but the moral tone adopted by the supporters of eugenics began to seem very hollow when the world saw what was going on in Nazi Germany under the name of "race hygiene." It was one thing to try to eliminate some "defective" genes—and quite another to identify the Aryan race as the chosen people and to try to purify it by eliminating anyone who did not fit the accepted standards. The Nazis' racist ideology, with all its horrific consequences for the Jews of Europe, owed at least as much to the idealist models of history and racial conflict as it did to Darwinism. But it was also the case that the Nazis took over a strong eugenics policy that had been established before they came to power, and that had been designed to eliminate people with serious mental, and even physical, disabilities. In the years following World War II, the new biological Darwinism combined with a wave of revulsion against the Nazi atrocities to undermine the credibility of "scientific" racism and eugenics.

It would be wrong to pretend that there have not been occasional efforts to revive the determinist position. One example that is relevant to a study of Darwinism is the so-called "anthropology of aggression." In 1924 the South African anatomist Raymond Dart discovered a fossil hominid which he called *Australopithecus africanus*. This had had a small-brain yet appeared to have walked upright, and Dart was convinced that it represented the earliest member of the human family tree. In effect, it confirmed Darwin's predictions that the human race had originated in Africa, and that the key evolutionary breakthrough had been the adoption of an upright posture as a means of walking on the open plains. Dart was ignored at first by the scientific community, which was convinced that big-brain human ancestors would be discovered in Asia. He was finally vindicated in the 1940s, when new fossil discoveries confirmed the significance of his original find. But Dart had by now developed very pessimistic views about the habits of our early ancestors. He pictured *Australopithecus* as a vicious predator and scavenger, prone to violence against its fellows and even cannibalism.

The possibility that such violent instincts could be programmed deep into the human mind was taken up by the popular writer Robert Ardrey. In books such as his *Territorial Imperative* (1966), Ardrey argued that humans are prone to violence because we still have the instincts of our hunting ancestors. Any social program that did not take this innate aggressive character into account would be doomed to failure. Such arguments inevitably lend themselves to the defense of both militarism and free enterprise, and Ardrey's position can easily been seen as a form of social Darwinism. Yet his argument actually owed little to the Darwinism of the new evolutionary biology. The claim that our ancestors were hunters, programmed with all the necessary aggressive instincts, could be defended by any evolutionist, Darwinian or non-Darwinian. The logic of the argument has been undermined not by challenging Darwinism, but by other anthropologists who have pointed out that the evidence that our distant ancestors ever gained most of their food from hunting is very slender. The macho image of "man the hunter" has been replaced with a less aggressive (but perhaps equally ideologically loaded) picture of "woman the gatherer."

There has been one genuine effort to apply the new Darwinism to human affairs: the science of sociobiology, which uses a strict Darwinism to explain the inherited instincts of social animals. The concept of the "selfish gene," introduced by the British biologist Richard Dawkins, expresses the logic of sociobiology: evolution will program organisms to behave in a way that is advantageous to the gene, not necessarily to the organism itself. Thus self-sacrificing behavior can be explained, because the individual's death may save the lives of genetic relatives who will transmit the genes to future generations. Sociobiologists believe that natural selection favors those genes that program organisms with apparently altruistic instincts—provided the altruism is expressed toward genetic relatives.

The American sociobiologist Edward O. Wilson has tried to apply the theory of the selfish gene to explain basic aspects of human nature. Certain kinds of behavior toward others, Wilson says, are "natural" in the sense that they express biologically implanted instincts arising from the evolutionary process. For example, it is natural for us to think first of our close relatives, and only secondarily of people less closely related to us. This seems to suggest that the family is a natural biological unit that must be respected in any form of social organization. More generally, selfish behavior (with appropriate qualifications for genetic relatives) is inevitable. Such implications are said by the movement's critics to represent an endorsement of right-wing policies as natural expressions of behavior that is imprinted in us as instincts.[16] It was inevitable that sociobiology would be branded as the new social Darwinism by left-wing thinkers and by social scientists determined to resist this new incursion of biology into human affairs.

This is not the place to enter into a discussion of the modern debates, but it should be noted that if sociobiology really is "social Darwinism," then it is the first real application of the strict logic of Darwin's selection theory to human affairs. Most earlier forms of social *evolutionism* were hardly Darwinian in the modern sense, and many of them, it turns out, were based on explicitly anti-Darwinian views. Spencer and many of the race theorists, for example, used Lamarckism even though they promoted the idea that struggle, in one form or another, played an important role in social development. To argue that the logic of Darwin's most important insight was not applied immediately to human affairs is not, however, to suggest that biology is somehow a pure science uncorrupted by ideology. The early understanding of "Darwinism" formed a seamless web in the public mind in which Darwin's use of the struggle for existence blended with Spencer's self-help progressionism. Darwin's theory was enmeshed in social debates from the time of its first publication, but for historical purposes we must be careful to note that it was the more general concept of evolution that played the most important role, not those aspects of Darwin's thinking that have turned out to be fruitful for modern biology.

# 5

## Darwin on Darwinism

If much late nineteenth-century evolutionism derived from Spencer and other writers who used what would now be called non-Darwinian concepts, one might argue that a truly representative text for the period would be one of Spencer's works, not one of Darwin's. Yet there is no doubt that Darwin's *Origin of Species* sparked off the great debate about evolution, and it remains one of the great pieces of scientific literature. It is thus to the *Origin* that we turn for an illustration of Darwinism, although many aspects of the theory it develops were not fully appreciated in Darwin's own time. An analysis of the argument and evidence provided in the *Origin* will allow us to appreciate those aspects of Darwin's thinking that are prized by modern biologists and will reveal why many of his contemporaries found it so difficult to see the selection theory as a complete explanation of evolution.

Most historians accept that it was the biogeographical discoveries made during the voyage of the *Beagle* that most determined Darwin's unique approach to evolutionism. He was aware from the start of the complex and unpredictable nature of the plant and animal communities to be found in any one location. He saw the interactions between species, and how dependent species were on their physical environment. When he turned to look for a mechanism to explain the bewildering variety, he wanted to find something that was based on natural laws but that could not be expected to yield a neat, orderly pattern of relationships. His decision to investigate the world of animal breeders put him in touch with an important area of practical biological knowledge—but it was an area that most biologists, with their emphasis on the "ideal type" of the species, would have studiously ignored.

By accepting the breeders' assertion that individual variation was essentially random or undirected, Darwin uncoupled the ultimate source of change from its directing agency. Just as the breeder could make almost anything he chose out of his animals, so could nature, depending on the pressures exerted on the population by its environment. Darwin's use of Thomas Malthus's population principle illustrates that he was also able to link in topics that were much discussed at the time, but not by other naturalists. The claim that the population always tended to outstrip the food supply threatened the traditional belief that species were created by a benevolent God. Malthus was exploited in the context of Darwin's preexisting belief that the relationship between a population and its environment was crucial. Darwin was interested in the process of evolution, and he looked at its products (the species of animals and plants) for clues as to how that process worked. He was not interested in reconstructing the whole history of life on earth and indeed thought that this was outside the scope of scientific possibility.

Darwin's contemporaries liked his descriptions of exotic locations and his emphasis on the interrelatedness of nature, but they remained convinced that behind the apparent diversity of the organic world there must be an order or pattern that could be discerned by the scientist. If nature was governed by law, its products must exhibit a rational order. The efforts made by many post-Darwinian biologists to reconstruct the history of life on earth were all based on the belief that science must be able to reveal the underlying pattern of evolution. Ideas about the nature of the process of evolution were thus determined by beliefs about the resulting pattern. Darwin reversed this relationship and thus escaped most of the temptation to see evolution as being based on a progressive trend or a tendency toward predetermined goals. His interest in the process of change was sophisticated enough to allow him to adopt a flexible policy that did not predict what the outcome must be. His efforts to study barnacles, and many plant structures, were designed not to reveal an underlying pattern in nature, but to exhibit the flexibility and unpredictability of the evolutionary process. History was vital to the understanding of the present state of nature. But it was history with all its unpredictable twists and turns, not a predetermined law of progress along a fixed hierarchy toward a preordained goal.

The *Origin of Species,* of course, contains the basic outline of the selection theory and of Darwin's overall view of the evolutionary process. Darwin's other best-known work is his *Journal of Researches into the Geology and Natural History of the Countries Visited during the Voyage of H.M.S.* Beagle. Originally published in 1845, this work is still frequently reprinted and is widely available, often under a title such as *Darwin's Voyage of a Naturalist.* Published long before the *Origin,* it nevertheless offers hints at some of Darwin's ideas on evolution, as well as illustrates the vast collection of information he acquired as he developed his ideas. Less well known and hardly ever reprinted, yet of crucial importance for understanding Darwin's views on human affairs, is his *Descent of Man* of 1871. The following sections of this chapter provide a guide to the most important sections of these three works.

## The Voyage of the *Beagle*

The *Journal of Researches* provides a lively account of Darwin's voyage on the *Beagle*—a classic example of Europe's growing interest in, and control over, the rest of the world.[1] It also describes the discoveries that Darwin made in and around South America, including his work on the Galápagos Islands. Darwin's early interest in geology is apparent, as he uses the evidence for the gradual uplift of the Andes mountains to support Lyell's uniformitarian theory. Anyone wishing to understand the lines of thought that led Darwin toward evolutionism should certainly read this book. However, historians now realize that the *Journal of Researches* was very much a product of Darwin's interests in the years following the voyage, when he was already developing his theory. Some of his earlier interests that actually took up much of his time on the voyage, for example his work on invertebrates, are mentioned only briefly.

Darwin was already an experienced amateur naturalist when he was invited by Captain Robert Fitzroy to accompany him as ship's naturalist aboard HMS *Beagle* on her second voyage to chart the coast of South America. The *Beagle* set out from England on 27 December 1831, and although Darwin was racked by seasickness, he was able to describe some of the mid-Atlantic islands visited by the ship on the way to South America. In February 1832 they reached

Bahia in Brazil, and for the first time Darwin saw a tropical rain forest: "The elegance of the grasses, the novelty of the parasitical plants, the beauty of the flowers, the glossy green of the foliage, but above all the general luxuriance of the vegetation, filled me with admiration" (*Journal*, p. 9). Darwin then spent several months in the vicinity of Rio de Janeiro (*Journal*, chapter 2). There he encountered a slave-owning society, and he recorded his horror of the institution of slavery and the cruel treatment that its victims were forced to endure. (Darwin's family had been prominent in the antislavery movement in Britain.)

In July 1832 the *Beagle* sailed to Montevideo, and Darwin was to spend most of the next two years traveling inland while the ship charted the coastal waters. He traveled across the pampas, or plains (chapters 3 and 8), and became quite used to living the life of a gaucho (a cowboy). He made an important discovery there: a new species of rhea, a flightless bird similar to the African ostrich. The first specimen of the new species had been killed and mostly eaten before Darwin realized its significance, but the remains were sent home and subsequently named *Rhea darwinii* in his honor. The newly discovered species had its main territory to the south of the common rhea, and over the next few years Darwin would puzzle over the relationship between the two species. He eventually realized that they must, in effect, be competing to occupy the territory at the boundary between their main ranges, and this prompted him to question the traditional view that each species is perfectly adapted to a single climate. He observed a far more obvious case of competition for territory when he encountered the troops of Argentina's General Rosas engaged in a war of extermination against the native Indians.

Throughout South America, Darwin discovered fossils (chapters 5 and 7). There was the extinct ground sloth, *Megatherium;* also fossil armadillos and a giant rodent, *Toxodon.* Darwin's comments in the *Journal* show how he reflected on these discoveries in the years following the journey. The fact that many of the fossils were related to living South American types helped him to establish the "law of succession of types." Each continent had its own characteristic forms of animals, that continued in a recognizable manner through geological time. For the mature Darwin, this would be a sign that they were linked by evolution, and that there was no single ladder of development that all life must necessarily ascend. Once a group of animals was trapped on an isolated continent, all their descendants would exhibit a clear relationship to the parent forms. Some of the fossils also helped to link now-distinct groups, confirming that evolution was a divergent rather than a linear process. The *Toxodon* combined rodent characteristics with those of pachyderms, leading Darwin to comment: "How wonderfully are the different Orders, at the present time so well separated, blended together in different points of the structure of the Toxodon!" (*Journal*, p. 59).

At the end of 1832 the *Beagle* passed through the Strait of Magellan and visited Tierra del Fuego (chapter 10). The ship carried three natives who had been taken from the archipelago by Captain Fitzroy on his previous visit, educated in England, and were now being returned in the hope that they could civilize the rest of their people. This was Darwin's opportunity to see native people living in what was then called a state of utmost savagery. He recorded his disgust at their dirty, brutal appearance and at the harshness of the lives they led—he noted that they would eat their old women before they would eat the dogs, when food got scarce. The three educated natives soon returned to the traditional life-style of their people, and a European missionary who landed with them had to be taken off in fear of his life. This episode forced Darwin to think very carefully about the relationship between civilized nations and the "savages" that were increasingly being seen as social equivalents of the Europeans' stone age ancestors. How had people so apparently brutal been given the trappings of civilization, and why did they then prefer to return to the harsh existence of their tribe? The relationship between the cultural and the biological differences between races would, as we have seen, plague anthropology through the rest of the century.

In 1834 the *Beagle* sailed around to the west coast of South America. On 20 February Darwin was ashore when he experienced a strong earthquake, and a few days later the ship sailed into the port of Concepción, Chile, to find that the town had been devastated (chapter 14). Darwin noted decaying shellfish now stranded several feet above the new shoreline, showing that the earthquake had elevated the land in relation to the level of the sea. On the nearby slopes of the Andes mountains he found beds of shells at considerable elevations, suggesting that the mountain range had been elevated not by a single great catastrophe, but by a series of normal earthquakes over a vast period of time. By the following year, Darwin was a complete convert to Charles Lyell's uniformitarian geology, according to which the earth's surface was shaped solely by forces that could still be seen in operation. All physical changes were slow and comparatively gradual, paving the way for a theory of equally gradual change in the organic world.

On 15 February 1835 the *Beagle* reached the Galápagos Islands, which lie on the equator several hundred miles out into the Pacific from the South American mainland (chapter 17). The islands are volcanic and had been formed quite recently by geological standards. Darwin was excited by the unusual animal life, including the giant tortoises and the birds, especially the finches. The *Journal* hints strongly at the radical conclusions he would later draw from the geographical diversity of the island species: "Hence, both in space and time, we seem to be brought somewhat nearer to that great fact, that mystery of mysteries—the first appearance of new beings on this earth" (*Journal*, p. 276).

The fact that the animals were different on each island was first brought to his attention in the case of the tortoises, when he was told that the natives could distinguish which island they came from by the shape of their shell. But it was the finches that Darwin used as the classic illustration of the effect that was to convert him to evolutionism. There were many different finches on the various islands—an expert ornithologist would later classify them into 13 species—each adapted to a different way of life. Some had thick beaks for breaking up seeds, other had thin beaks for extracting insects from crevices.

Darwin stated the implications of this diversity in a group of related species quite explicitly, if only as a hypothesis: "seeing this gradation and diversity of structure in one small, intimately related group of birds, one might really fancy that from an original paucity of birds in this archipelago, one species had been taken and modified for different ends" (*Journal*, pp. 276-77). On his return to England, Darwin would begin a search for a mechanism of evolution that would modify these isolated populations by adapting them to their local environment. We now know that the conclusions hinted at in the *Journal* were very much a product of hindsight, because Darwin only noticed the geographical diversity among the finches just before the *Beagle* was about to sail, too late for him to undertake careful studies on the spot. He had to reconstruct the distribution of the finches from memory, although he had better records for the mockingbirds and may actually have recognized the geographical variation first among this group.

The *Beagle* now sailed home via Tahiti, New Zealand, Australia, and the Cape of Good Hope. Darwin saw coral reefs in the Pacific and developed a theory to explain how they were formed: by a gradual subsidence of the land into the sea (chapter 20). The coral animals cannot live at a great depth, so only slow subsidence could explain how they keep building up to the surface. A gradually sinking island would thus build up an ever-thickening girdle of coral reefs around it, until eventually the island itself would have disappeared, leaving only a shallow coral atoll. This theory became generally accepted after Darwin's return and helped him to make his reputation as a geologist long before he became known for his work in natural history. It provided yet more confirmation of Lyell's gradualist geology.

Darwin offers brief comments on the other places visited on the way home, but he was not given the opportunity for lengthy study. The *Beagle* finally reached England on 2 October 1836, nearly five years after setting out. Darwin would spend the next few years living in London, writing up his experiences, developing his reputation as a geologist, and secretly gathering information that might bear on the question of the transmutation of species.

## The *Origin of Species*

Darwin himself described the *Origin of Species* as "one long argument," and it is important that readers are primed to understand the sequence of points developed in the case for evolution. The book also contains a wealth of information and detail on many aspects of natural history, illustrating the scope of Darwin's vast program of research into the evidence for and implications of his theory. The *Origin* is a densely written book, but it is relatively free of technical jargon and can still be read by anyone with only a smattering of knowledge of natural history. It thus remains one of the last great works in the literature of science to aim its argument directly at the educated general public as well as the scientific community. The first edition, published in 1859, is available in facsimile reprint, but most popular reprints are of the sixth edition of 1872. Although this edition includes the final version of Darwin's argument, it is in some respects less satisfactory because the central argument is obscured by Darwin's responses to the many objections raised against the theory. This last edition contains a whole new chapter answering the critics (chapter 7). The following discussion focuses on the first edition, representing the form in which Darwin's argument first reached the public.[2]

The full title of Darwin's book is *On the Origin of Species, or, The Preservation of Favoured Races in the Struggle for Life.* The "Introduction" primes the reader so that he or she will understand the point of the book. Darwin stresses that the central problem of evolution is the origin of adaptive structures, by no means an obvious point, given that many naturalists believed that all species possess nonadaptive characters. Critics have, in fact, complained that the *Origin of Species* is really about the origin of adaptations, not the origin of species at all. There is some truth in this, although Darwin does address the question of how one species can split into several "daughter" species. But certainly natural selection is primarily an explanation of adaptive evolution, and Darwin is anxious to stress that he has a new insight on this question. This explains why the *Origin* begins with a description of the selection mechanism, not with the general case for evolution. Darwin knew that many of his readers would be familiar with the earlier debates on the transmutation of species that were widely regarded as having become bogged down because, with Lamarckism discredited, no one could think of a scientifically plausible explanation of how the process might work. Darwin was hoping to break this log jam by presenting his new theory, and he was successful in reopening the case for evolution, even though natural selection itself remained highly controversial.

The first five chapters of the book thus form a solid description of the selection mechanism. Darwin begins in chapter 1 with "Variation under Domestication," because he was convinced that the best way of getting his readers to understand natural selection would be to confront them with the process of artificial selection and the evidence for the vast changes produced in species by human intervention. Once this is understood, Darwin believed the reader would more readily accept that there might be an analogous process in nature that could have similar or even greater effects. Darwin's purpose was to convince the reader that a species is not a collection of identical organisms all cast in the same mold: each organism is an individual with slight but unique character differences. Breeders exploit this variability to produce many different races, by selecting those individuals who just happen to have the characteristics they desire. Darwin insists that there is no automatic tendency for a breed to revert back to some "natural" type that is basic to the species. This aspect of his thinking is crucial to modern Darwinism, even though genetics has given us a different method of explaining variation. To most mid-nineteenth-century naturalists, a species was a thing-in-itself with a fixed essence; variations from the norm were trivial and accidental. To suppose that such variations were crucial for the future evolution of the species was to imply that evolution had no built-in law of development—hence Herschel's charge that natural selection was the law of higgledy-piggledy.

Note that almost from the start Darwin introduces his nongenetical theory of variation and heredity: new characteristics are induced by environmental disturbance of the reproductive process. Such disturbances are often hereditary, and thus form the raw material for natural selection; they are the random or apparently purposeless differences that exist among all members of a population. In some cases, however, the effect of the environment may be to induce useful modifications, and this allows Darwin to accept a subsidiary role for Lamarckism. These points are developed at greater length in chapter 5 of the *Origin*, where Darwin insists that variations are *caused* even though we speak of them as being "random." They are random only in the sense that most of them are produced without reference to the needs of the species.

In chapter 2, "Variation under Nature," Darwin moves on to argue that at least some random variation must exist within wild species. He concedes that there may not be as much as under domestication, because artificial conditions will stimulate the disturbances to growth that generate variations. But no natural environment is ever totally stable, so Darwin is able to suggest that even wild species will exhibit small amounts of random individual variation. He suggests that the local varieties which exist within many species must have been built up out of this individual variation by a process that would, if taken further, produce distinct species. This is supported by the fact that naturalists often disagree over whether a particular local form is merely a variety of another species, or a species in its own right. "[A] well-marked variety may justly be called an incipient species" (*Origin*, p. 52).

Chapter 3 introduces the "Struggle for Existence," which Darwin uses as the agent in place of the human animal breeder in natural selection. The struggle arises from the tendency of all species to over reproduce: more individuals are born than can survive, and so there must be competition to determine which will stay alive long enough to reproduce. The origin of this insight is clear: "It is the doctrine of Malthus applied with manifold force to the whole animal and vegetable kingdoms" (*Origin*, p. 63). The competition occurs between the individuals of the same species, and many factors determine survival—ability to cope with shortage of food, predators, and so on.

Darwin emphasizes the complex web of what we would now call ecological relationships between species, and many of his contemporaries were enthralled by his account of these interactions. His image of the "tangled bank," with its interwoven network of different plants, each dependent on the others, may have helped his readers to appreciate the complexity of natural relationships. Too much should not be made of this, however, since there is little evidence that even the scientists were ready to take this complexity seriously. It was not until the 1890s that detailed study of what we now call ecology began. Most evolutionism before then was taken up with the study of dead organisms dissected in laboratories, and the study of fossils, in the hope of elucidating their structural relationships. This is another example of how Darwin's emphasis on the complexity and unpredictability of nature was not acceptable to his contemporaries.

Darwin admits that behind its facade of apparent beauty, nature is a scene of constant struggle. Yet he is anxious not to create too harsh an image lest this upset his readers' preconceived ideas about universal harmony. He implies that his theory modifies but does not overthrow the traditional view that nature has been created by a wise and benevolent God: "We may console ourselves with the full belief, that the war of nature is not incessant, that no fear is felt, that death is generally prompt, and that the vigorous, the healthy, and the happy survive and multiply" (*Origin*, p. 79).

Chapter 4 outlines the complete theory of natural selection: given the production of random individual variations, and the rigid dependence of the wild organism on its environment, there must occasionally arise slight modifications of structure that are useful in the struggle for existence. "If such do occur, can we doubt (remembering that many more individuals are born than can possibly survive) that individuals having any advantage, however slight, over others, would have the best chance of surviving and of procreating their kind?" (*Origin*, p. 80). These useful characteristics will be passed on to the next generation, while any useless or harmful ones will be eliminated. Darwin stresses that nature's ability to scrutinize variations will be far more acute than that of the human animal breeder, and will act over much longer periods. He ends by once again trying to minimize the apparent harshness of his proposed mechanism, proposing that natural selection will always act to improve the organism's ability to cope with its environment, and will thus, in a sense, promote the improvement of life on earth.

Survival is not, of course, enough to make natural selection work: the successful organisms must also reproduce, and Darwin was well aware that reproductive success might not depend on survival alone. In chapter 4 he provides a brief description of his theory of "sexual selection" (discussed at much greater length in the *Descent of Man*). A trait that gives an individual a better chance of breeding, because it gives an edge in the competition for mates, will be enhanced even if it is of no adaptive value. Thus the bright colors of many male birds are explained as a result of these colors being attractive to females during the species' mating behavior. The males with the brighter colors have always mated more successfully— even if their color makes them more conspicuous to predators.

In a world governed by natural selection, extinction is always a possibility when a species is confronted by a rival—perhaps an invader of its home territory—that is better adapted than itself. Darwin was aware that many species must have gone extinct, with only a few surviving and diversifying to replace them. He believed that species are constantly tending to specialize, to become better adapted to a particular way of life that avoids competition with potential rivals. For this reason, Darwin saw evolution as an ever-branching tree or bush, in which some branches are halted by extinction while others go on to divide yet again. The diagram he gives at this point to illustrate the process is significant: unlike many evolutionary trees, Darwin's has no central trunk. There is no main line of evolution, because each branch is "doing its own thing," and cannot be evaluated by standards derived from another. In theory, at least, Darwin does not see the human race as the goal toward which all evolution has been working.

Chapter 5 describes in more detail Darwin's pregenetical ideas about the origin of variations, and is perhaps of less interest to most modern readers. With chapter 6 he moves into a new phase of the argument, in which he responds to the various objections that he anticipates will be raised against his theory. The theory of divergent evolution is invoked to explain why we do not see one species blending imperceptibly into another. Opponents of evolution, then and now, often claim that if one form evolves gradually into another, we should see instead of distinct species a continuum of forms with no gaps by which they can be separated. This argument ignores the branching character of evolution. Darwin claims that species are produced by a gradual, continuous process of modification, but when branching occurs the less-specialized intermediates are eliminated as the branches move further apart. Thus "species come to be tolerably well-defined objects, and do not at any one period present an inextricable chaos of varying and intermediate links" (*Origin*, p. 177). To give an example not used by Darwin, humans and chimpanzees have evolved from a common ancestor that was apelike, but was certainly not a chimpanzee; the two modern species lie at the ends of two diverging lines leading away from the common ancestor.

Another problem is the formation of unusual and highly complex structures. Darwin anticipated that his readers would at first find it difficult to believe that a functioning wing, or an eye, could have been created by the kind of process he is suggesting. Darwin points out, however, that there are many intermediate structures that can help us to see how the most perfect form was developed. Flying squirrels show that an organism can make good use of a less-than-perfect flying ability, illustrating the kind of "halfway house" through which the ancestors of bats and birds must have passed. The human eye was a particular problem, since it was widely regarded as being so perfect a structure that only supernatural design could have created it. But again Darwin notes that there are many different degrees of visual ability in the animal kingdom, and even quite simple eyes are useful to the species that possess them. He thus proposes that from a very simple beginning, perhaps just a spot of light-sensitive tissue, the various steps in the ascent toward the human eye could each have appeared and been favored by natural selection. Darwin admits that if one could find a complex organ for which no intermediate steps were conceivable, then his theory would be undermined, but he does not believe that any such organs can be found.

The argument from design is, in any case, unable to explain examples of species that do not use their physical adaptations, such as web-footed geese that do not live in water. Such cases are to be expected, however, if species are sometimes forced to change their habitat, and then they take some time to adapt to their new way of life. Darwin also addresses the issue of structures of little or no importance. If natural selection works only by favoring adaptations, then species should not possess any nonadaptive characteristics, but Darwin points out that naturalists may not always be aware of a structure's importance and may thus exaggerate the number of nonadaptive traits. The giraffe's tail seems too small to be any use, but it serves as a flyswatter in conditions where freedom from insect pests may be vital for a species' survival.

Darwin was particularly concerned to show that his theory could account for the evolution of instinctive behavior patterns. In chapter 7 he explains his belief that instincts are subject to variation just like any physical characteristic, and may thus be affected by natural selection. The Lamarckians explained instincts as the end product of learned habits that had gradually become hereditary, but Darwin notes that this theory does not account for the instincts of neuter insects, such as worker ants and bees. The selection theory can tackle this problem, because the process can be imagined to act on families as well as individuals: insects with a tendency to produce useful neuter types among their offspring would succeed as the founders of colonies. (Here Darwin slips into what is now know as group selection; modern sociobiology has shown that the case of neuter insects can be dealt with on the basis of individual selection at the level of the gene.)

Darwin turns, in chapter 8, to what was often seen as a fatal objection to evolutionism, the question of hybridism. The classic definition of a species is based on the assumption that individuals from two distinct species cannot reproduce. Local varieties of the same species, by contrast, and breeds produced by human interference, are fully interfertile. This distinction was held to mark the boundary between natural productions (varieties) and supernaturally created species. Darwin argues that the distinction is not as sharp as everyone had assumed. The mule is a cross between the horse and the ass, but is normally ignored in this context because it is infertile. Darwin shows, however, that there are varying degrees of sterility in the hybrids of different species, with hybrids of closely related species often having some fertility. The allegedly sharp distinction between species and varieties, is in fact, so unclear that it confirms Darwin's own belief that well-marked varieties are incipient species. As two originally very closely related forms gradually diverge, their ability to interbreed steadily diminishes until at last they are unable to reproduce at all.

In chapter 9 Darwin tackles what has always been one of the main objections to evolutionism, the fossil record. Far from showing a gradual evolution of life on earth, the record often seems to indicate that new forms were introduced suddenly. The title of the chapter indicates Darwin's line of defense: "On the Imperfection of the Geological Record." Following the geologist Charles Lyell, he argues that fossils are only produced in unusual circumstances, so most species will have lived and become extinct without leaving any trace in the rocks. The fossil record is like a book that has had most of its pages destroyed, he explains—we should not expect it to read like a smooth narrative of the history of life on earth. On the question of the apparently sudden appearance of life on earth at the beginning of the Cambrian period (Darwin calls it the Silurian in the first edition; the nomenclature has since changed), he admits a problem. Clearly, if his theory is true, complex organisms could not have appeared suddenly. Darwin postulates a long initial period of evolution in which no fossils were laid down, perhaps because the organisms had only soft bodies, or because the continents were then in different positions. Darwin could only hope that Precambrian fossils would eventually be discovered, although it was not until the twentieth century that this would come to pass.

Chapter 10 continues with the theme of the fossil record, but Darwin now moves on to provide positive arguments to support his theory. He notes that once the imperfection of the record is allowed for, the pattern of development matches exactly with what his theory would predict. The first members of any new group to appear are usually generalized forms that seem to combine characteristics now separated in distinct members of the group. In the course of time the group divides into a number of more distinct lines of descent that seem to lead, by increasing specialization, toward the modern members of the group. Thus pigs and camels are seen to belong to a single group, because they can be united by means of less specialized fossils. The pattern is exactly what one would expect if evolution were a branching process brought about by adaptive specialization.

Chapters 11 and 12 deal with the kind of evidence that had originally led Darwin to accept evolution, the geographical distribution of species. The present distribution of species makes no sense in terms of the old theory of divine creation, since often very different forms are found living in identical climates. Each continent has its own characteristic forms of life, indicating that geographical barriers have exerted a powerful influence on the development of life. Darwin argues that a successful new form will spread out from its original home to occupy as much territory as possible, and will be halted by major barriers such as the oceans. He also discusses the process by which isolated islands such as the Galápagos get their populations by accidental transportation from the nearest mainland; then the island species generally specialize into unique forms. By this time of writing the *Origin of Species,* however, Darwin had abandoned the belief that speciation (the division of one species into many) occurs only when a single original population gets broken up across geographical barriers. He thought that on a large landmass, the pressure of specialization would force sections of a single population to evolve in different directions and would thus pull the species apart. This view was to prove very difficult for his followers to uphold.

In the penultimate chapter (chapter 13) Darwin uses his theory to explain the relationships that naturalists see between species and that they use to classify them. At one time, naturalists sought a "natural" system of classification in which the relationships corresponded to those in the divine plan of creation. Darwin argues that the true natural system is a genealogical one based on a process of branching evolution. Those species that have only recently diverged from a common ancestor are grouped most closely together; more remote relationships correspond to more distant branching points on the tree of life. Darwin notes that in some cases one can observe relationships only in the embryos of species whose adult forms are widely different. The embryo represents the species in a less specialized, less modified state, and thus offers a clue to its distant ancestry. Note, however, that Darwin does not endorse the simplest version of the recapitulation theory, in which the human embryo passes through the whole sequence of vertebrate evolution. In his model of branching evolution one cannot see the development of the human embryo as a speeded-up representation of the whole history of life on earth. Life has not climbed a ladder up to humankind, leaving the "lower" animals as the rungs on the ladder beneath, because each modern type is the end product of its own distinct branch of evolution.

Finally, Darwin refers to the case of rudimentary organs that are too small to serve any purpose. These made no sense on the basis of creationism—why should a wise and benevolent God burden species with useless characteristics? But they make perfect sense in an evolutionary context, for changing circumstances may force a species to change its behavior so that a once-useful organ no longer serves any purpose. Evolution will not wipe that organ out instantaneously, but it will gradually diminish its size, and we should expect many species still to possess rudiments of the structures they once used in the past.

In his concluding chapter Darwin sums up the general outline of his new approach to nature and expresses the hope that younger naturalists will appreciate its value and throw off the shackles of creationism, which, in fact, explains nothing. If they do, he says, there will be a revolution in natural history. A famous sentence, "Light will be thrown on the origin of man and his history" (*Origin*, p. 488), provides the only absolutely clear indication that Darwin wishes to extend his theory into this controversial area. To counter this disturbing implication, however, Darwin ends with an attempt to show that his theory can still be reconciled with traditional beliefs, saying that it is better to imagine the Creator governing the world by law than by unpredictable miracles. Again he returns to the theme of progress, suggesting that "as natural selection works by and for the good of each being, all corporeal and mental endowments will tend to progress towards perfection" (*Origin*, pp. 488–89).

Nature is not arbitrary and purposeless, Darwin asserts, because it has been designed as a system that will allow progress to occur in the long run, despite the complexity of the system. "Thus, from the war of nature, from famine and death, the most exalted object of which we are capable of conceiving, namely, the production of the higher animals, directly follows. There is a grandeur in this view of life, with its several powers, having been originally breathed into a few forms or into one; and that, whilst this planet has gone cycling on according to the fixed law of gravity, from so simple a beginning endless forms most beautiful and most wonderful have been, and are being, evolved" (*Origin*, p. 490). These concluding words stress the idea of evolution as a process governed by law, with an ultimate purpose built into it. The last word is, incidentally, the only instance in which Darwin uses any derivative of the term *evolution* in the book.

The *Origin of Species* was widely read, and the logic of its argument must have been appreciated by its readers. Surely all were struck by the breadth of Darwin's knowledge of the complexities of nature. Modern biologists are constantly amazed at the ways in which he anticipated important ideas and concepts that have only become widely discussed in the twentieth century. Yet we have seen that, in Darwin's own time, few were prepared to accept that his argument was strong enough to justify the belief that natural selection explains the whole process of evolution. The general case for evolution was indeed reassessed, and in the end Darwin's position would triumph. But few naturalists could accept that a process driven only by random variation and the demands of the local environment could have generated the living things we see around us. What the majority of naturalists wanted was a law-bound system of evolution that revealed rather more clearly the element of ultimate purpose that even Darwin conceded. The system of interacting natural processes that Darwin invoked was so complex that the end products were, to all intents and purposes, unpredictable and haphazard (higgledy-piggledy, to use Herschel's phrase). If there really was a purposeful or progressive trend in nature, most nineteenth-century thinkers preferred to believe that it was built more securely into the laws governing life itself. It was not enough to have progress as a mere by-product of a process whose immediate goal was local adaptation.

## *The Descent of Man*

*The Descent of Man* is a much less readily available book. The first edition (1871) was published in two volumes; the second edition (1874), published in one volume, is easier to find in libraries. One chapter was repositioned between in the second edition—chapter 4 of the first edition became chapter 2 of the second. The following outline is based on the more accessible second edition.[3]

Much of the book, which is written in three parts, is taken up with the question of sexual selection (its full title is *The Descent of Man and Selection in Relation to Sex*). In the long second part, Darwin makes the general case for sexual selection. And in the third part of the book Darwin applies his theory of sexual selection to explain certain human characteristics that he thought were nonadaptive, and that thus could not have been produced by natural selection or Lamarckism. The following analysis concentrates on the first part, in which Darwin makes the general case for human evolution.

Darwin begins in chapter 1 of part I by stressing that the structure of the human body provides clear evidence of our affinity with the animals. For this T. H. Huxley had paved the way with his *Man's Place in Nature*, which had discussed the anatomical resemblances between human and apes. Darwin emphasizes that the human embryo is, at an early stage, virtually indistinguishable from that of any other mammal. He provides many illustrations of rudimentary structures in the human body that only make sense when seen as relics of an animal ancestry. One example is what has come to be known as "Darwin's point" on the human ear, a vestige of the pointed ears of our animal ancestors.

Chapter 2 is in some respects the most important in the book, and it contains insights that would not be appreciated until the twentieth century. Darwin here tries to explain how the human race might have evolved from a generalized ape ancestor (contrary to popular misconception, the evolutionists did not believe that we had evolved from a modern ape such as the gorilla). Natural selection is the basic mechanism, of course, although in this case Darwin was rather more willing to concede a role for the inheritance of acquired characteristics. He insists that the human race is subject to variation and to the Malthusian tendency to increase its numbers beyond the level that can be supported by the supply of food, both of which are essential for natural selection to work.

The real question, though, is why a process of adaptive change should have given us characteristics that differ so markedly from the apes that have evolved from the same common ancestor. Darwin notes that the human race is the most dominant animal on earth, and he accepts that this is mainly due to our intellect and our ability to communicate via language. But he also stresses the importance of our hands as a physical adaptation to manipulate the environment. The human hand is much more dexterous than that of an ape, and Darwin believes this is due to the fact that it is no longer used for locomotion in the trees.

Unlike most of his contemporaries, Darwin did not see an increase in intelligence as the driving force of evolution. It would be easy to assume that more intelligence is always beneficial—but in that case, why didn't the apes keep up with humans? Darwin emphasizes instead the significance of those bodily characteristics by which humans do differ from the apes, especially our hands and our upright posture. He argues that the crucial development came when our ancestors left the forests in which the apes had always lived, and to which the modern apes remain confined. The earliest human adapted to walking upright because this was the best method of locomotion on the open plains. As a consequence, their hands were freed to explore the environment and to begin using primitive tools. Although he never states it explicitly, Darwin assumes that humans evolved because of a change of habitat and the consequent adaptive modifications. The human brain is a by-product of adaptation, not the goal of a universal progressive trend. Once our human ancestors began to walk upright, there would have been a number of consequential changes in the human body. Their teeth would have diminished in size as they began to make weapons to use instead, for hunting and defense. As their intelligence increased, the brain would have necessarily increased in size, and this would have altered the shape of the skull.

In chapter 3 Darwin seeks to compare the mental powers of animals and humans. Clearly it was in his own interest to magnify the level of animal mentality, to minimize the gap to be bridged. Darwin concedes that there is an immense gulf between the lowest savage and the highest ape, yet he insists that the higher animals possess at least the rudiments of all the human mental functions: "My object . . . is to shew that there is no fundamental difference between man and the higher mammals in their mental faculties" (*Descent*, p. 66). Darwin argues that the range of mental powers possessed by animals is often underestimated: the lower animals perform complex actions by instinct, which requires no intelligence, but the higher animals have to work out how to deal with new situations by their own intelligence. He also stresses the range of abilities displayed by the various human races, thereby betraying his acceptance of the common racial stereotypes of his time. His object here was to imply that the gulf between the highest animal and lowest human levels of intelligence was small when compared to the great range on either side.

Darwin also tries to argue that the rudiments of all the human emotions are found in the animals, including wonder and curiosity. Modern psychologists think that Darwin greatly exaggerated the mental powers of animals; he adopted an uncritical anthropomorphism that allowed him to see traces of even the highest functions— including morality and spirituality—in animals. He sought to explain the origin of language through the instinctive copying of natural sounds and held that the development of this faculty would go hand in hand with the growth of intelligence. Where he felt that there was little or no evidence for a particular mental function in animals, he questioned whether it was all that well developed in the "lowest" humans. Are primitive savages really-self-conscious in the same way as Europeans? he asks. According to John Lubbock, the archaeologist who distinguished the New from the Old Stone Age, the first stone tools were probably produced accidentally—we should not ascribe forethought to the primitives who first used them.

Chapters 4 and 5 deal with the moral powers explicitly and present them as an essential feature of any social animal endowed with intelligence. Darwin argues "that any animal whatsoever, endowed with well-marked social instincts, the parental and filial affections being here included, would inevitably acquire a moral sense or conscience, as soon as its intellectual powers had become as well, or nearly as well developed as in man" (*Descent*, p. 98). Significantly, however, he notes that if the social instincts of this hypothetical animal were different from ours, then its moral sense would be different too, a clear challenge to the idea of absolute moral standards.

For animals to live in groups, Darwin points out that evolution must provide them with instincts promoting cooperative behavior. He gives many examples of this kind of behavior, including cases in which animals risk their own lives to save those of other members of their group. He also records the example of a monkey who saved the life of a friendly zookeeper by distracting a fierce baboon that was attacking him. Darwin was never sure if these altruistic instincts had been formed by natural selection or by Lamarckism. Natural selection has certainly tended to eliminate those individuals with antisocial tendencies. Here Darwin also accepts the possibility of group selection: those groups with less cooperative behavior would be eliminated when in conflict with those whose members cooperated more effectively. He also invokes Lamarckism by arguing that social habits may eventually turn into inherited instincts: humans learn the benefits of cooperating with other members of our group, and perhaps the resulting habits eventually become so ingrained that they are inherited.

"Savages," Darwin believed, illustrate the most primitive form of moral behavior, in which the greatest good is defined in terms of benefit to the tribe, not benefit to the human race as a whole. Thus primitive peoples appeared to be immoral when judged by the standards of European explorers. Only with the development of civilization have the social instincts been generalized into universal moral laws. Darwin insists that even under civilized conditions there is a constant tendency for antisocial individuals to be eliminated. He worries, however, about the tendency for the unfit to be protected in a civilized society, thus anticipating the concerns of the eugenics movement.

Chapter 6 returns to the question of humankind's links to the animal kingdom and the thesis that our closest relatives are the great apes. Darwin theorized that the birthplace of the human species was probably southern Africa, where both the chimpanzee and the gorilla still live. This was a theory that modern paleoanthropologists would regard as fully confirmed, but the vast majority of Darwin's contemporaries thought that central Asia, with its harsher and hence more stimulating climate, was a more probable location. The chapter concludes with a list of the stages of vertebrate evolution leading up to humankind.

In chapter 7, the final chapter of part I, Darwin turns to the races of mankind. He considers the arguments advanced by the polygenists, those physical anthropologists who claimed that the various races are distinct biological species, and finds them inadequate in the face of the observed facts that the races can all interbreed successfully and that they grade insensibly into one another where they meet. He challenges the view of the German anthropologist Karl Vogt, that the various human races had evolved from different ape ancestors, pointing out that this would imply a most unlikely degree of convergence between several distinct evolutionary lines. Darwin certainly shared the racist opinions of his time, being quite convinced that "savages" are intellectually and morally inferior to the white race. He even quoted figures intended to show that colored races have smaller brains than whites. When faced with an invading "higher" race, Darwin believed that primitive groups tend to go into a decline that often leads to extinction (even when there is no direct hostility from the invaders). Yet he would not go along with those who wished to sever the links between the races altogether. For Darwin, all of humankind originated from the same primitive human type, and all share a basic common humanity.

Part II of *The Descent of Man*—by far the longest section—deals with sexual selection, a topic that is outside the scope of this discussion. Darwin launched into the topic because he was convinced that many of the differences between the races could not be explained in terms of adaptation to the various areas of the world into which the human species has migrated. When he returns to human evolution in part III, it is to invoke sexual selection as the mechanism that explains these differences. The various races have different standards of male and female beauty, Darwin argues, and these influence the choice of mates in a way that enhances the desired characteristics. Even the color differences between the races may be due to this cause, he suggests. As a more general example, Darwin believed that the loss of body hair by all races was a result of sexual selection. Few biologists or anthropologists of the late nineteenth century accepted sexual selection as an explanation of human characteristics; indeed the basic concept of sexual selection attracted little attention until the mid-twentieth century.

In his conclusion to *The Descent of Man*, Darwin concedes that many will find it difficult to accept the idea that humans have evolved from the lower animals. Yet all educated people had come to accept that civilization had developed from savagery, so all should have been be prepared to accept that we have evolved from people no better than the natives of Tierra del Fuego whom Darwin himself had encountered on the voyage of the *Beagle*—naked, bedaubed with paint, their expression wild, startled, and distrustful. Darwin shared the prevailing belief of his time that modern "savages" were equivalent to the stone age ancestors of the so-called higher races. He contrasts the barbarous customs of these primitive humans with the examples he had cited of self-sacrificing behavior in animals, and expresses the view that he would as soon be descended from the animals as from savages. The fact that we have progressed so far gives us hope for the future, but he argues that we should never forget that we bear the stamp of our animal origins.

*The Descent of Man*, like the *Origin of Species*, is both a classic document of its age and a surprising anticipation of modern ideas. There can be little doubt that Darwin took on many of the preconceptions of his time about the primitive nature of savages and the inevitability of progress leading up to the higher character of the white race. Yet in other respects his approach to the question of human origins and his ability to examine simple progressionism raise Darwin above the level of his contemporaries. More than any other Victorian evolutionist, Darwin asked the kinds of questions about how the apes and humans diverged that a modern paleoanthropologist would find significant. Some of his answers, too, were much closer to modern thinking than those offered by his contemporaries, especially his predictions that Africa would turn out to be the site of human origins and that our earliest ancestors would have walked upright before they began to have bigger brains. Darwin emerges as a figure who was both immersed in the culture of his time and capable of transcending it when exploring the logic of his own, very radical, theory of evolution. The unique character of Darwin's writings thus helps us to understand how others at the time could accept the general idea of evolution, but not his particular explanation of it.

# Notes and References

1. FROM DARWIN TO MODERN DARWINISM

1. A good example of this "Darwin centered" approach is Loren Eiseley's *Darwin's Century: Evolution and the Men Who Discovered It* (New York: Doubleday, 1959); for a more recent expression of the same view see Ernst Mayr's *The Growth of Biological Thought* (Cambridge, Mass: Harvard University Press, 1982).

2. For a development of these historiographical points, see Peter J. Bowler, *The Non-Darwinian Revolution: Reinterpreting a Historical Myth* (Baltimore: Johns Hopkins University Press, 1987).

3. On the significance of these later developments, see Peter J. Bowler, *The Mendelian Revolution: The Emergence of Hereditarian Concepts in Modern Science and Society* (London: Athlone; Baltimore: Johns Hopkins University Press, 1989).

4. See Edward O. Wilson, *Sociobiology: The New Synthesis* (Cambridge, Mass.: Harvard University Press, 1975); on the resulting debate, see Arthur L. Caplan, ed., *The Sociobiology Debate* (New York: Harper & Row, 1978).

5. The claim that scientific knowledge is socially constructed is developed in, for instance, Steve Woolgar, *Science: The Very Idea* (Chichester, England: Ellis Horwood / London: Tavistock, 1988) and in James Robert Brown, *The Rational and the Social* (London: Routledge, 1989).

6. On these rival models of history, see Peter J. Bowler, *The Invention of Progress: The Victorians and the Past* (Oxford: Basil Blackwell, 1989).

## 2. DARWINISM IN SCIENCE

1. On geology in the early nineteenth century, see Charles Gillispie, *Genesis and Geology* (Cambridge, Mass.: Harvard University Press, 1951). For further works on many of the topics developed in this chapter see the material outlined in the following Bibliographic Essay.

2. There is a modern reprint available of Lyell's *Principles of Geology* (in three volumes) with an introduction by M. J. S. Rudwick (Chicago: University of Chicago Press, 1991).

3. There is an English translation of Lamarck's *Zoological Philosophy* (London: John Murray, 1914); see also Richard W. Burckhardt, Jr., *The Spirit of System: Lamarck and Evolutionary Biology* (Cambridge, Mass.: Harvard University Press, 1977).

4. See Frank Sulloway, "Darwin and His Finches: The Evolution of a Legend," *Journal of the History of Biology* 15 (1982): 1–54.

5. Darwin's sketch and essay are reprinted along with Alfred Russel Wallace's 1858 paper in Darwin and Wallace, *Evolution by Natural Selection* (Cambridge: Cambridge University Press, 1958).

6. Parts of Darwin's unfinished book have been published in Robert C. Stauffer, ed., *Charles Darwin's Natural Selection: being the Second Part of his Big Species Book Written between 1856 and 1858,* (Cambridge: Cambridge University Press, 1975).

7. See J. V. Jensen, "Return to the Huxley-Wilberforce Debate," *British Journal for the History of Science* 21 (1988): 161–80. On Huxley's view of evolution see Adrian Desmond, *Archetypes and Ancestors: Palaeontology in Victorian London* (London: Blond and Briggs, 1982; Chicago: University of Chicago Press, 1990), and Mario Di Gregorio, *T. H. Huxley's Place in Natural Science* (New Haven: Yale University Press, 1982).

8. Richard Owen, *On the Anatomy of Vertebrates,* (London: Longmans, Green, 1866-68, 3 vols.), III, p. 808.

9. E. Ray Lankester, *Degeneration: A Chapter in Darwinism* (London: Macmillan, 1880).

10. E. W. MacBride, *A Textbook of Embryology,* vol. 1, *Invertebrates* (London: Macmillan, 1914), 663. See Peter J. Bowler, "E. W. MacBride's Lamarckian Eugenics," *Annals of Science* 41 (1984): 245–60.

11. Matthew's 1914 paper is reprinted in his *Climate and Evolution* (New York: New York Academy of Sciences, 1939); see Ronald Rainger, "Just before Simpson: William Diller Matthew and Evolution," *Proceedings of the American Philosophical Society* 130 (1986): 453–74.

12. Fleeming Jenkin, "The Origin of Species," *North British Review* 46 (1867): 277–318; this is reprinted along with other reviews in David

L. Hull, ed., *Darwin and His Critics* (Cambridge, Mass.: Harvard University Press, 1973).

13. See Arthur Koestler, *The Case of the Midwife Toad* (London: Hutchinson, 1971); Koestler's interpretation of these events is unreliable, however.

### 3. DARWINISM, RELIGION, AND MORALITY

1. Jacques Barzun, *Darwin, Marx, Wagner: Critique of a Heritage* (Garden City, N.Y.: Doubleday, 1958); Gertrude Himmelfarb, *Darwin and the Darwinian Revolution* (New York: Norton, 1959).

2. W. F. Moneypenney and G. E. Buckle, *The Life of Benjamin Disraeli*, rev. ed., (London: John Murray, 1929, 2 vols.). II, p. 108.

3. The myth of the conflict between science and religion is challenged, for instance, in James R. Moore, *The Post-Darwinian Controversies: A Study of the Protestant Struggle to Come to Terms with Darwinism in Britain and America, 1870–1900* (Cambridge: Cambridge University Press, 1979).

4. See David Livingstone, *Darwin's Forgotten Defenders: The Encounter between Evangelical Theology and Evolutionary Thought* (Edinburgh: Scottish Universities Press; Grand Rapids, Mich.: Eerdmans, 1987).

5. Asa Gray, *Darwiniana: Essays and Reviews Pertaining to Darwinism* (New York: Appleton, 1876), 147–48.

6. Darwin, *The Variation of Animals and Plants under Domestication,* 2nd ed., London: John Murray, 1882, 2 vols.), II, p. 428.

7. John Frederick William Herschel, *Physical Geography* (Edinburgh: A. & C. Black, 1861), 12. On the "higgledy-piggledy" comment, see Darwin's letter to Lyell, 12 December 1859, in Francis, Darwin, ed., *The Life and Letters of Charles Darwin,* (London: John Murray, 1887, 3 vols.), II, p. 241.

8. Richard Owen, *On the Anatomy of Vertebrates,* (London: Longmans, Green, 1866–68), III, p. 808.

9. John Dewey, *The Influence of Darwin on Philosophy and Other Essays on Contemporary Thought* (1910; reprint, Bloomington: Indiana University Press, 1965), 12.

10. Samuel Butler, *Essays on Life, Art and Science* (London: R. A. Streatfield, 1908), 308.

11. George Bernard Shaw, *Back to Methuselah: A Metabiological Pentateuch* (London: Constable, 1921), liv.

12. Darwin, *The Descent of Man and Selection in Relation to Sex* (London: John Murray, 1871, 2 vols.), I, pp. 138–45. Chapter 2, in which this

insight occurs, was repositioned as chapter 4 in the second edition (London: John Murray, 1874), see pp. 49–53.

## 4. SOCIAL DARWINISM

1. Richard Hofstadter, *Social Darwinism in American Thought*, revised ed. (New York: George Braziller, 1959).

2. See Robert C. Bannister, *Social Darwinism: Science and Myth in Anglo-American Social Thought* (Philadelphia: Temple University Press, 1979).

3. See for instance Robert M. Young, *Darwin's Metaphor: Nature's Place in Victorian Culture* (Cambridge: Cambridge University Press, 1985), and John C. Greene, *Science, Ideology and World View* (Berkeley and Los Angeles: University of California Press, 1981).

4. See Donald C. Bellomy, " 'Social Darwinism' Revisited," *Perspectives in American History*, n.s. 1 (1984): 1–129.

5. On the rival models of history see Peter J. Bowler, *The Invention of Progress: The Victorians and the Past* (Oxford: Basil Blackwell, 1989).

6. This point is emphasized in Robert J. Richards, *Darwin and the Emergence of Evolutionary Theories of Mind and Behavior* (Chicago: University of Chicago Press, 1987).

7. Herbert Spencer, *Social Statics: Or the Conditions Essential to Human Happiness Specified* (London: John Chapman, 1851), 378–79.

8. See Spencer, *Principles of Biology*, (London: Williams and Norgate, 1864, 2 vols.), I, p. 444.

9. On the link between Lankester's and Wells's views on degeneration, see Peter J. Bowler, "Holding Your Head up High: Degeneration and Othogenesis in Theories of Human Evolution," in James R. Moore, ed., *History, Humanity and Evolution: Essays for John C. Greene* (Cambridge, Cambridge University Press, 1989), 329–53.

10. Arthur Koestler, *The Case of the Midwife Toad* (London: Hutchinson, 1971), p. 82. On MacBride's views on race, see Peter J. Bowler, "E. W. MacBride's Lamarckian Eugenics," *Annals of Science* 41 (1984): 245–60.

11. Karl Pearson, *The Grammar of Science*, 2nd ed. (London: A. & C. Black, 1900), p. 369.

12. Vernon Kellogg, *Headquarters Nights: A Record of Conversations and Experiences at the Headquarters of the German Army in France and Belgium* (Boston: Atlantic Monthly Press, 1917).

13. W. J. Sollas, *Ancient Hunters and Their Modern Representatives* (London: Macmillan, 1911), 383. On changing ideas about human origins, see Peter J. Bowler, *Theories of Human Evolution: A Century of De-*

*bate, 1844–1944* (Baltimore: Johns Hopkins University Press; Oxford: Basil Blackwell, 1986).

14. For the classic account of the exposure of the Piltdown fraud, see J. S. Weiner, *The Piltdown Hoax* (Oxford: Oxford University Press, 1955); since then, many unsatisfactory efforts have been made to identify the culprit. On the fossils' role in the theoretical debate, see Bowler, *Theories of Human Evolution*.

15. Arthur Keith, *The Antiquity of Man* (London: Williams and Norgate, 1915), 136.

16. See Arthur L. Caplan, ed., *The Sociobiology Debate* (New York: Harper and Row, 1978).

## 5. DARWIN ON DARWINISM

1. Darwin, *Journal of Researches into the Natural History and Geology of the Countries Visited during the Voyage of H.M.S. Beagle* (1845; reprint, London: Routledge, 1891). Hereafter cited in text.

2. Darwin, *On the Origin of Species; or, The Preservation of Favoured Races in the Struggle for Life* (London: John Murray, 1859; reprint, Cambridge, Mass.: Harvard University Press, 1964). Hereafter cited in text.

3. Darwin, *The Descent of Man and Selection in Relation to Sex*, 2nd ed. (London: John Murray, 1874). Hereafter cited in text.

# Bibliographic Essay

The study of Darwinism is one of the most active areas in the history of science, with a wide range of books and technical articles available—and more being added every year. The following bibliographic guide concentrates mainly on books and journals that should be available in any good library. The unwary reader must be warned that recent historical studies have overturned or questioned many of the traditional interpretations of Darwin's work and its impact. The older literature (and some of the modern popular accounts) often must be treated with caution.

Darwinism is an immensely controversial topic, and it is easy for historians to reflect their own opinions in their interpretation of the past. Some classic studies, including Gertrude Himmelfarb's *Darwin and the Darwinian Revolution* (New York: Norton, 1959), are marred by a distrust of Darwin's theory that seems to extend to a dislike of all things Darwinian, including Darwin himself. The writings of the scientist-turned-historian may just as easily be biased in the opposite direction, seeing Darwin as the hero of discovery who established the framework of modern evolutionism almost single-handed. Such studies are unlikely to look with sympathy on the modern tendency to interpret Darwin's cultural and social environment as having been an important factor in the development and popularization of his theory. These conflicting interpretations may confuse the reader who expects to find simple answers to his or her questions about Darwin, but they are a sign that the debate over the origins of evolutionism is both active and still relevant to modern affairs.

There are a number of general studies of the history of evolutionism. The best of the older surveys is John C. Greene's *The Death of Adam: Evolution and Its Impact on Western Thought* (Ames: Iowa State University Press, 1959). Written for a more popular audience, and far less reliable, is Loren Eiseley, *Darwin's Century: Evolution and the Men Who Discovered It* (New York: Doubleday, 1959). A more recent survey written by one of the architects of modern Darwinism in biology is Ernst Mayr, *The Growth of Biological Thought: Diversity, Evolution and Inheritance* (Cambridge, Mass.: Harvard University Press, 1982). The most comprehensive guide to the subject, including an extensive bibliography, is Peter J. Bowler, *Evolution: The History of an Idea*, rev. ed. (Berkeley and Los Angeles: University of California Press, 1989). For an explicit challenge to the Darwin-centered view of the history of evolutionism, see Peter J. Bowler, *The Non-Darwinian Revolution: Reinterpreting a Historical Myth* (Baltimore: Johns Hopkins University Press, 1988) and *The Invention of Progress: The Victorians and the Past* (Oxford: Basil Blackwell, 1989).

A classic survey of pre-Darwinian evolutionism is Bentley Glass, Owsei Temkin, and William L. Strauss, eds., *Forerunners of Darwin: 1745–1859* (Baltimore: Johns Hopkins University Press, 1959). On the early nineteenth century, see Charles C. Gillipsie, *Genesis and Geology: A Study of the Relations of Scientific Thought, Natural Theology, and Social Opinion in Great Britain, 1790–1850* (Cambridge, Mass.: Harvard University Press, 1951). In the past decade, historians have realized that the concept of evolution was widely debated in the early nineteenth century, thus upsetting many traditional ideas about the period. Two important studies are Tobey Appel, *The Cuvier-Geoffroy Debate: French Biology in the Decades before Darwin*, (New York and Oxford: Oxford University Press, 1987, and Adrian Desmond, *The Politics of Evolution: Morphology, Medicine and Reform in Radical London* (Chicago: University of Chicago Press, 1989). There have also been important changes in our understanding of the history of geology in this period; see for instance Stephen Jay Gould, *Time's Arrow, Time's Cycle: Myth and Metaphor in the Discovery of Geological Time* (Cambridge, Mass.: Harvard University Press, 1987), and Anthony Hallam, *Great Geological Controversies* (Oxford: Oxford University Press, 1983).

Adrian Desmond's *Archetypes and Ancestors: Palaeontology in Victorian London, 1850–1875* (London: Blond and Briggs; Chicago:

University of Chicago Press, 1982) straddles the Darwinian revolution and offers important insights into the social framework of the debate. Also valuable on the fossil evidence for evolution are Martin J. S. Rudwick, *The Meaning of Fossils: Episodes in the History of Paleontology* (New York: Science History Publications, 1972), and Peter J. Bowler, *Fossils and Progress: Paleontology and the Idea of Progressive Evolution in the Nineteenth Century* (New York: Science History Publications, 1976).

Turning to Darwin himself, his *Journal of Researches* and *Origin of Species* are widely available in reprints, although the latter are usually of the sixth edition; the first edition is reprinted by Harvard University Press (1964). Darwin's earliest accounts of his theory are available in Charles Darwin and Alfred Russel Wallace, *Evolution by Natural Selection* (Cambridge: Cambridge University Press, 1958), which also includes Wallace's 1858 paper. There is now a major project under way to publish all of Darwin's private papers; see Paul H. Barrett et al., eds., *Charles Darwin's Theoretical Notebooks (1836–1844)* (Cambridge: Cambridge University Press, 1987), and Frederick Burckhardt and Sydney Smith, eds., *The Correspondence of Charles Darwin* (Cambridge: Cambridge University Press, 1984–; 8 vols. published to date).

There are many biographies of Darwin, although some of the more popular ones are unreliable as guides to how historians now interpret his work. One of the classics is Gavin De Beer, *Charles Darwin* (London: Nelson, 1963). A modern biography that stresses the social context of Darwin's work is Adrian Desmond and James R. Moore, *Darwin* (London: Michael Joseph, 1991); for a survey of modern thinking on Darwin, see Peter J. Bowler, *Charles Darwin: The Man and His Influence* (Oxford: Basil Blackwell, 1990). For a massive compilation of recent Darwin scholarship see David Kohn, ed., *The Darwinian Heritage* (Princeton: Princeton University Press, 1985). Michael Ruse, *The Darwinian Revolution: Science Red in Tooth and Claw* (Chicago: University of Chicago Press, 1979), offers a good survey of the origins of the theory and the subsequent debate.

Specialized studies of the development of Darwin's ideas include Howard E. Gruber, *Darwin on Man: A Psychological Study of Scientific Creativity* (New York: E. P. Dutton, 1974); Dov Ospovat, *The Development of Darwin's Theory: Natural History, Natural Theology and Natural Selection, 1838–1859* (Cambridge: Cambridge University Press, 1981); and Edward Manier, *The Young Darwin and His Cultural Circle* (Dordrecht, Netherlands: D. Reidel, 1978). There is an immense amount of technical literature on this issue; it is surveyed in David Oldroyd, "How Did Darwin Arrive at His Theory?" *History of Science* 22 (1984): 325–74. On Darwin's scientific work, including the broader applications of his theory in biology, see Michael Ghiselin, *The Triumph of the Darwinian Method* (Berkeley and Los Angeles: University of California Press, 1969).

On the scientific debate surrounding the publication of the theory, see David Hull, *Darwin and His Critics: The Reception of Darwin's Theory of Evolution by the Scientific Community* (Cambridge, Mass.: Harvard University Press, 1973), which includes reprints of many reviews of the *Origin of Species*. On the debate in different countries, see Thomas F. Glick, ed., *The Comparative Reception of Darwinism* (Austin: University of Texas Press, 1974). Darwin's response to his critics is outlined in Peter J. Vorzimmer, *Charles Darwin: The Years of Controversy* (Philadelphia: Temple University Press, 1970). On the rise of anti-Darwinian evolution theories, see Peter J. Bowler, *The Eclipse of Darwinism: Anti-Darwinian Evolution Theories in the Decades around 1900* (Baltimore: Johns Hopkins University Press, 1983).

A valuable guide to the popular reception of the theory in Britain is Alvar Ellegard, *Darwin and the General Reader: The Reception of Darwin's Theory of Evolution in the British Periodical Press, 1859–1872* (Göteborg: Acta Universitatis Gothenburgensis, 1958; reprint, Chicago: University of Chicago Press, 1990). See also Cynthia E. Russett, *Darwin in America: The Intellectual Response* (San Francisco: W. H. Freeman, 1976). On the religious debate, the best survey is James R. Moore, *The Post-Darwinian Controversies: A Study of the Protestant Struggle to Come to Terms with Darwin in Britain and America, 1870–1900* (Cambridge: Cambridge University Press, 1979); see also David Livingstone, *Darwin's Forgotten Defenders: The Encounter between Evangelical Theology and Evolutionary Thought* (Edinburgh: Scottish Universities Press; Grand Rapids, Mich.: Eerdmans, 1987). The various essays in Robert M. Young's *Darwin's Metaphor: Nature's Place in Victorian Culture* (Cambridge: Cambridge University Press, 1985) explore the social context of Darwin's theory and its reception.

On the evolution of the mental and moral powers, see Robert J. Richards, *Darwin and the Emergence of Evolutionary Theories of Mind and Behavior* (Chicago: University of Chicago Press, 1987). More specifically on human origins, see Peter J. Bowler, *Theories of Human Evolution: A Century of Debate, 1844–1944* (Baltimore: Johns Hopkins University Press; Oxford: Basil Blackwell, 1986). The background of cultural evolutionism is outlined in J. W. Burrow, *Evolution and Society: A Study in Victorian Social Theory* (Cambridge: Cambridge University Press, 1966), and more widely in George W. Stocking, *Victorian Anthropology* (New York: Free Press, 1987). See also Thomas R. Trautman, *Lewis Henry Morgan and the Invention of Kinship* (Berkeley and Los Angeles: University of California Press, 1987), and Bowler, *The Invention of Progress* (cited above).

On the race question, see John S. Haller, *Outcasts from Evolution: Scientific Attitudes of Racial Inferiority, 1859–1900* (Urbana: University of Illinois Press, 1975); Nancy Stepan, *The Idea of Race in Science: Great Britain, 1800–1960* (London: Macmillan, 1982); and Stephen Jay Gould, *The Mismeasure of Man* (New York: Norton, 1981). On the breakdown of the evolutionary paradigm in the social sciences, see George W. Stocking, *Race, Culture and Evolution* (New York: Free Press, 1968), and Hamilton Cravens, *The Triumph of Evolution: American Scientists and the Heredity-Environment Controversy, 1900–1941* (Philadelphia: University of Pennsylvania Press, 1978).

The classic account of social Darwinism is Richard Hofstadter, *Social Darwinism in American Thought*, rev. ed. (New York: George Braziller, 1959). This is criticized in Robert C. Bannister, *Social Darwinism: Science and Myth in Anglo-American Social Thought* (Phildelphia: Temple University Press, 1979). See also Greta Jones, *Social Darwinism and English Thought* (London: Harvester, 1980). A comprehensive survey of eugenics is Daniel Kevles, *In the Name of Eugenics: Genetics and the Uses of Human Heredity* (New York: Knopf, 1985); see also Mark H. Haller, *Eugenics: Hereditarian Attitudes in American Thought* (New Brunswick, N.J.: Rutgers University Press, 1963); G. R. Searle, *Eugenics and Politics in Britain, 1900–1914* (Leiden, Netherlands: Noordhoff International Publishing, 1976); and Cravens, *The Triumph of Evolution* (cited above).

On the emergence of modern biological Darwinism, see William B. Provine, *The Origins of Theoretical Population Genetics* (Chicago: University of Chicago Press, 1971); Ernst Mayr and William B. Provine, eds., *The Evolutionary Synthesis: Perspectives on the Unification of Biology* (Cambridge, Mass.: Harvard University Press, 1980); and Marjorie Grene, ed., *Dimensions of Darwinism: Themes and Counterthemes in Twentieth-Century Evolutionary Theory* (Cambridge: Cambridge University Press, 1983). Peter J. Bowler, *The Mendelian Revolution: The Emergence of Hereditarian Concepts in Modern Science and Society* (London: Athlone; Baltimore: Johns Hopkins University

Press, 1989), stresses the extent of the conceptual changes associated with the rise of genetics.

Differing views among working biologists on the implications of the mid–twentieth-century Darwinian synthesis are apparent from Julian Huxley's introduction to Pierre Teilhard de Chardin, *The Phenomenon of Man* (London: Collins, 1959); for the radical materialist perspective, see George Gaylord Simpson, *The Meaning of Evolution* (New Haven: Yale University Press, 1949) and *This View of Life: The World of an Evolutionist* (New York: Harcourt Brace, 1963). Arthur Koestler's *The Ghost in the Machine* (New York: Macmillan, 1967) is a critique of the kind of materialism represented by Simpson's view of nature.

On sociobiology, see Edward O. Wilson, *Sociobiology: The New Synthesis* (Cambridge, Mass.: Harvard University Press, 1975), and Richard Dawkins, *The Selfish Gene*, 2nd ed. (Oxford: Oxford University Press, 1990). On the human implications of sociobiology see Wilson's *On Human Nature* (Cambridge, Mass.: Harvard University Press, 1978); for a survey of the resulting controversy see Arthur L. Caplan, ed., *The Sociobiology Debate* (New York: Harper & Row, 1978). For an account of the whole range of modern controversies over Darwinism, inside and outside science, see Michael Ruse, *Darwinism Defended: A Guide to the Evolution Controversies* (Reading, Mass.: Addison-Wesley, 1982), and the concluding chapter of Bowler, *Evolution: The History of an Idea* (cited above). Good accounts by biologists of the implications of modern evolutionism include Stephen Jay Gould, *Wonderful Life: The Burgess Shale and the Nature of History* (New York: Norton, 1989), and Richard Dawkins, *The Blind Watchmaker: Why the Evidence of Evolution Reveals a Universe without Design* (New York: Norton, 1987). Michael Ruse, *Taking Darwin Seriously: A Naturalistic Approach to Philosophy* (Oxford: Basil Blackwell, 1986), argues that even the sociobiologists have not fully thrown off the legacy of progressionism.

# Index

# The Author

Peter J. Bowler is Professor of the History and Philosophy of Science at The Queen's University of Belfast, Northern Ireland. He is the author of numerous books on the history of evolution theory and its implications, and of a survey history of the environmental sciences.

# The Editor

Michael S. Roth is the Hartley Burr Alexander Professor of Humanities at Scripps College and professor of history at the Claremont Graduate School. He is the author of *Psycho-Analysis as History: Negation and Freedom in Freud* (1987) and *Knowing and History: Appropriations of Hegel in 20th-Century France* (1988), both published by Cornell University Press. He is currently writing about contemporary strategies for representing the past in the humanities and about conceptualizations of memory disorders in the nineteenth century.